# WILD RICE

S. G. Aiken

P. F. Lee

D. Punter

J. M. Stewart

Published by
NC Press Limited
Toronto

in cooperation with
Agriculture Canada
and the
Canadian Government Publishing Centre
Supply and Services Canada

1988

Agriculture Canada Publication 1830

Catalogue number: A53-1393/1988E

© Minister of Supply and Services Canada – 1988

No part of this publication may be reproduced, stored in a retrieval system, or transmitted, in any form or by any means, electronic, mechanical, photocopying, recording or otherwise, without the prior written permission of NC Press Limited.

*Cover illustration:* Wayne Yerxa of the Couchiching Indian Reserve, Fort Frances, Ontario. The original painting is part of the private collection of Dr. Don and Mrs. Frances Leishman, Thunder Bay, Ontario.

---

**Canadian Cataloguing in Publication Data**

Wild rice in Canada

Co-published by Agriculture Canada.
Bibliography: p.
Includes index.
ISBN 1-55021-027-0

1. Wild rice – Canada.   I. Aiken, S. G.
II. Canada.   Agriculture Canada.

SB191.W55W54  1988        633.1'8'0971

C88-094529-X

---

We would like to thank the Ontario Arts Council and the Canada Council for their assistance in the production of this book.

**New Canada Publications,** a division of NC Press Limited, Box 4010, Station A, Toronto, Ontario, Canada, M5W 1H8.

Printed and bound in Canada

# To William (Bill) G. Dore

It is with great pleasure that the authors, under the auspices of the Biosystematics Research Centre, Agriculture Canada, dedicate this publication to Dr. W. G. Dore, the Canadian authority on wild rice during the past 25 years. The original intent was to publish a second edition of Dore's publication on wild rice (Agric. Can. Publ. 1393) with minor alterations. Some portions of the original publication are unchanged, particularly much of the introduction, the description of the wild rice plant, and the section on early methods of harvesting and processing. Information in the taxonomic section has been reorganized, most of the observations having been made by Bill Dore. Since the issue of the first edition, however, much new information has become available. Advances have been made in our understanding of wild rice: environmental interactions; habitat requirements; competition; population dynamics; diseases; and the domestication and cultivation of this cereal as a crop. The present authors accept responsibility for the new sections.

Wild Rice in Canada

## Contents

**Introduction**    7
    Acknowledgments    9

### I. Wild Rice Plant    11
    Roots    11
    Stems    13
    Leaves    14
    Flowers    16
    Grains    17
    Grain germination and viability    19

### II. A Taxonomic Review of the Genus, Species, and Varieties    21
    Genus *Zizania*    21
    Taxonomic categories within the genus *Zizania*    23
    Key to the annual varieties of wild rice    26
    *Zizania aquatica*    30
    *Zizania palustris*    32
    *Zizania texana*    36
    *Zizania latifolia*    38

### III. Wild Rice Habitat    39
    Water depth    39
    Water chemistry    41
    Type of soil    42
    Competition with other plants    43
    Competition within wild rice populations    47

### IV. Management of Natural Stands    48
    Lake selection    48
    Seeding of potential sites    49
    Water depth management    53
    Soil management    53
    Weed control    54
    Thinning requirements    57
    Straw removal    58

## V. Diseases and Pests    59
    Diseases    59
        *Brown spot*    59
        *Leaf sheath and stem rot*    61
        *Anthracnose*    62
        *Leaf blotch*    62
        *Smut*    62
        *Ergot*    63
        *Minor fungal pathogens*    71
        *Bacterial and viral diseases*    71
    Pests    72
        *Invertebrates*    72
        *Vertebrates*    78

## VI. Harvesting and Processing the Grain    80
    Harvesting in natural stands    80
    Early methods of harvesting and processing    81
    Current methods of harvesting and processing    84
    Food value and uses    92

## VII. The Wild Rice Industry    95
    Government involvement    95
        *Alberta*    96
        *Saskatchewan*    97
        *Manitoba*    97
        *Ontario*    100
        *New Brunswick*    103
        *Nova Scotia*    103
        *Prince Edward Island*    103
        *Other provinces*    103
    Impact of paddy culture on the wild rice industry    104
    Organization of the wild rice industry    105
    Industry quality control    106
    Outlook    106

*References*    109
*Additional Reading*    115
*Glossary*    121
*Index*    125

# Introduction

Wild rice (*Zizania* L.) is an annual aquatic grass that grows in shallow lakes and rivers throughout eastern and north central North America. It is best known as a culinary delicacy and as an export product, in addition to being a visible component of Indian culture. Of the wild grasses in Canada, it is the only one that grows from seed each year and produces a grain of sufficient size to be used extensively as food by people. The other cereals, wheat, oats, barley, rye, and corn, are "tame" grasses in the sense that they are completely dependent on the farmer's care and attention for survival in Canada's climate. These crop plants had their origin in foreign lands, and for many centuries they have been so modified by selection and cultivation that they are now quite unlike their wild progenitors. In contrast, wild rice is essentially the same today as it was when the first explorers found certain Indian tribes of the interior of North America using it as a main food.

The natural "wild rice bowl" where the grain has been used for centuries as a food, extends over an area west of Lake Superior to southern Manitoba and into the adjacent states of Wisconsin and Minnesota (Jenks 1901). Several rivers and lakes in the north central states and in Canada have received their names from the presence of stands of wild rice. Rice Lake near Peterborough, Ont., is perhaps the largest and best known of these lakes in Canada. Another lake, at 50°31'N, 93°31'W, northeast of Kenora, is called Zizania Lake.

Wild rice is found mainly in shallow water along the shores of rivers and streams, where stands are dense and continuous. In lakes, where it is usually less abundant, stands are generally concentrated near the inlet and outlet, where the current is more or less constant. Its lighter green color usually distinguishes wild rice from bordering stands of cattails and other shallow-water plants. Natural wild rice stands are economically valuable and ecologically, as well as aesthetically, desirable.

Historically, wild rice was an important food item in the diet of Indians west of Lake Superior (Jenks 1901). The tribes of the Algonquin and Sioux linguistic groups were the traditional wild rice gatherers of the "wild rice bowl." On the basis of preserved wild rice grains found during archaeological excavation of burial mounds, Johnson (1969) believes that wild rice was an important Indian food for at least 1000 years. The dependence of the Indian culture on the wild rice crop was such that when it failed, famine resulted. Although the reasons for failures were many and varied in the historical records, high water levels drowning out the wild rice were invariably linked to poor yields.

The many names for wild rice reflect the influence of Indians, explorers, fur traders, and early settlers on the use of this food: Indian rice, Canadian rice, water oats, water rice, and *manomin* (Densmore 1974). The last one was derived from an Indian name. The early French explorers called the plant *folle avoine*. Today direct translations, for example, *riz sauvage* and *zizanie*, are used in writings in French. However, *folle avoine* is also the common name for wild oats *(Avena fatua)*, and this dual use is confusing. Many producers of wild rice, as well as discerning naturalists and sportsmen who wish to be precise about the plant, use the name zizania. This is written without a capital letter and in ordinary print, but otherwise the word is identical with the Latin name of the genus to which our species of wild rice belong. Because such usage conforms with botanical principles of priority, zizania is clearly understood internationally.

In Agric. Can. Publ. 1393, *Wild-rice,* Dore explained that the hyphen in the name was used to reduce confusion in parts of the world where rice *(Oryza sativa)* is the chief cereal crop. All the native species of *Oryza* L. in Asia, as well as weedy strains of *O. sativa* that infest the fields in southern United States, are appropriately called wild rice, that is, rice that grows wild. The single word, wildrice, has been used to distinguish North American *Zizania* from Asian *Oryza* wild rice. Although this is useful if the subtlety is understood, the single-word form has not been widely accepted. Many recent publications, especially those from Minnesota, have used the two-word form, unhyphenated. This is consistent with the guidelines suggested in *Hortus Third* and by Alex et al. (1980).

Research has been initiated in Northwestern Ontario, Saskatchewan, and Alberta to provide the necessary background information before large-scale "farming" (aquaculture) of selected wild rice lakes can occur (Lee 1984). Stands of wild rice tolerate wide ranges in environmental conditions, for example, water depths (0.05–2.50 m), sediments (clay to peat), and latitudes (30° to 56°N). Because wild rice grows in a changing physical, chemical, and biological environment, the production of biomass is rarely predictable; consequently, good or bad years of wild rice production are considered normal. A key to stabilizing production would be to control the environmental factors causing fluctuations. This is one of the strategies used in modern agriculture, but its use is predicated upon our scientific understanding of those environmental factors whose interactions determine the success or failure of the yearly wild rice crop.

In order to stabilize environmental variables and control growing conditions, wild rice has been grown in paddies, particularly in the United States (Brooks 1981). Wild rice grown in such fields now accounts for about 90% of all rice marketed, but paddy production is relatively expensive and is still at the experimental stage in Canada. Here, many lakes suitable for growing wild rice are still being discovered and developed. In the process, interest in and the production of lake-grown wild rice crops is expanding. It is hoped that this publication will further this interest.

## Acknowledgments

The authors thank J. A. Percich for allowing them to use some of the colored photographs illustrating diseases of wild rice; C. E. Beddoe for taking Figures 2 and 6; S. J. Darbyshire for Figure 5 *d, e*; and M. Jomphe for detailed drawings. They also thank H. Schumer, North Central Experimental Station, Grand Rapids, Minnesota, and Dr. E. A. Oelke, M. McClellan, and the team at the University of Minnesota, St. Paul, for sharing enthusiastically information about work being done in Minnesota; and P. Sain, Manitoba Department of Natural Resources, Winnipeg, for information on wild rice in Ontario. They express their sincere thanks to many people involved in the production of this publication: M. A. Martin for typing the manuscript; B. M. Hilliker for extensive word processing help; P. M. Catling, S. J. Darbyshire, T. A. Steeves, and S. I. Warwick for reading the manuscript and suggesting improvements; and Sheilah Balchin, Dodie Archibald, and Sharon Rudnitski for their editorial assistance.

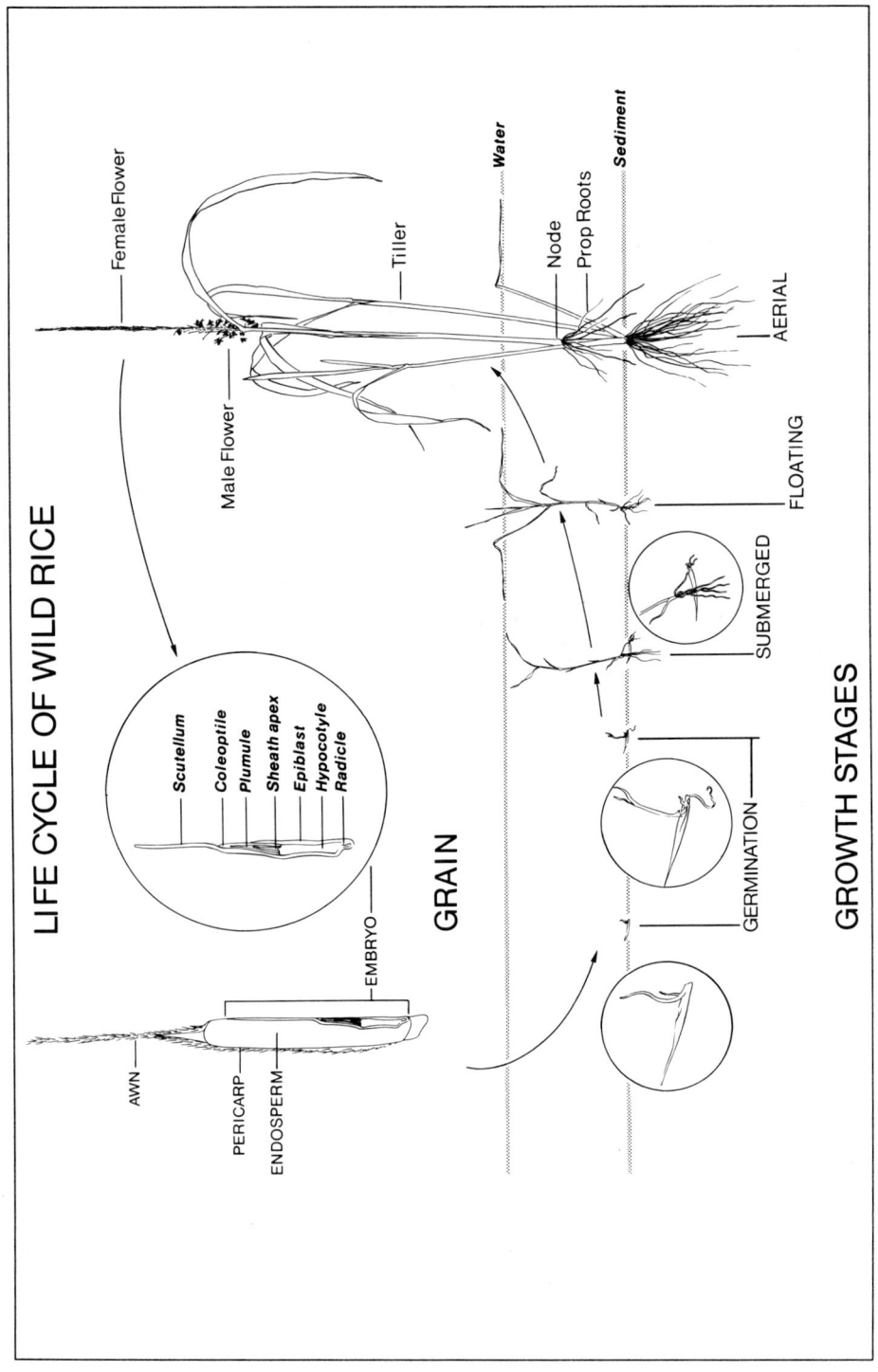

# I
# Wild Rice Plant

Figure 1 illustrates the life cycle of wild rice. Wild rice, like most other vascular plants, is composed of roots, stems, leaves, flowers, and fruits. To supplement the following description of these structures, morphological details are illustrated and described (Figs.1–6).

## Roots

The primary root is the first root to emerge from a germinating seed (Aiken 1986) (Fig. 2). It pushes through the seed covering close to the point where the seed was attached to the parent plant and persists for about a month. Soon after the primary root appears, a permanent adventitious root system develops from the first stem node and later from higher nodes. These roots are similar to the prop roots of corn; they grow into the mud diagonally and anchor the plant firmly against the lift force of waves and currents. The adventitious roots are straight, spongy, and light-colored or often rust-tinged from iron deposited on their surface. They have short, horizontal rootlets and no true root hairs, which makes wild rice unusual among grasses. The root system of a mature plant is shallow (up to 35 cm deep) with a lateral spread approximating the circumference of the aerial leaf coverage.

Fig. 1 *(opposite)* Life cycle of wild rice illustrating a longitudinal section of a wild rice grain and the germination, submerged, floating, and aerial stages of growth relative to water levels.

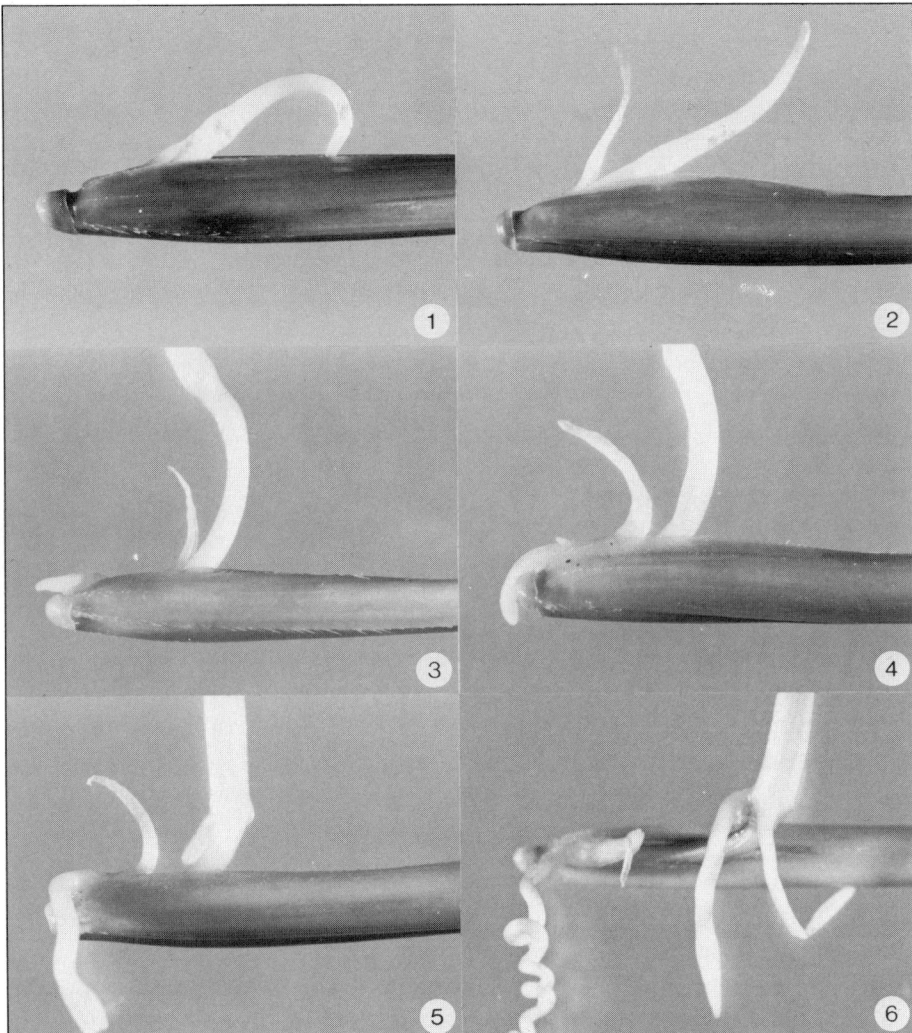

Fig. 2. Stages in the germination of a wild rice grain: *(1)* The epiblast has been pushed out through the lemma surrounding the grain by the seedling mesocotyl and coleoptile; the two structures are still closely associated. *(2)* The epiblast *(left)* has separated from the elongating mesocotyl *(center)*. *(3)* The root tip *(left)* is emerging through the seed covering close to where the seed was attached; the epiblast *(center)*; the elongated mesocotyl *(right)*, which in wild rice may be 1–6 cm long. The bulge toward the top of the photograph is the transition between the mesocotyl and the base of the coleoptile. *(4)* The elongating root *(left)* is curving toward the substrate. The intact lemma can be seen between the root and the epiblast. *(5)* A well-developed primary root *(left)*; secondary roots *(center)* beginning to develop from the base of the stem. *(6)* A curled primary root *(left)* showing curling that appeared to result from growing the seedling in a glass container of water (not known to occur in a mud substrate); the two secondary roots are developing rapidly to become the permanent rooting system *(center)*. Photographs of various grains, all magnified approximately 3×.

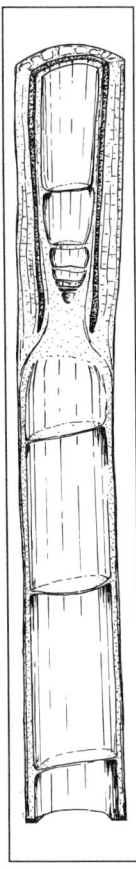

Fig. 3. Lengthwise section of stem through one node, showing parchment-like cross-partitions. Approximately 4.2×.

## Stems

The first shoot structure to appear is a sheath (coleoptile), that covers the first leaf (Fig. 2). The structure most commonly referred to as a stem, or culm, develops from near the base of the first leaf and elongates to become the supporting axis that bears the leaves and flowers. In wild rice, the embryo stem (mesocotyl) may elongate up to 6 cm to allow the seedling to emerge through overlying sediments; however if the seeds germinate on the sediment surface, very little basal elongation occurs.

The stems of grasses are surrounded by closely enveloping sheaths. Ordinarily the stems of wild rice are visible in the emergent stage and vary in height from 0.6 to 3.0 m (rarely 5.0 m) depending on the variety, plant density, competition, nutrient status, water depth, and other associated environmental factors. Stems in the middle of a wild rice stand are usually uniform in height, whereas the height of those near the margins of the stands is influenced by the depth of water. Branch stems (tillers) arise from the basal nodes and are usually shorter than the main stem. A typical wild rice stand in water 1 m deep with four plants per 900 cm$^2$ would have from three to five tillers per plant, but as many as 50 tillers per plant have been reported from stands in shallow water.

At the emergent stage, the culm is cylindrical with band-like joints and fine, felt-like hairs at the nodes. In cross section these nodes are a mass of continuous tissue. The stem section between the nodes is called the internode, and there are from three to five internodes in mature plants. The lower internodes are about 30 cm long and the upper ones can be as long as 90 cm. The internodes are smooth and are usually constant in number for each ecotype. Internally, the internodes are hollow, air-filled cavities that act as air locks and assist the buoyancy of the whole plant; hence, plants that are uprooted float to the surface and never regain a roothold. Within the cavities are thin, delicate, cross-partitions of remnant pith tissue that is porous and allows the diffusion of gases up and down the stem (Fig. 3). In aquatic plants, good aeration must extend to the roots, which grow in sediments that may be anaerobic (Sculthorpe 1967).

## Leaves

Following the emergence of the coleoptile, the next two or three leaves to appear are thin, pale green, and ribbon-like. They grow rapidly and are characterized by the absence of epicuticular wax.

Each succeeding leaf is larger than the preceding one. When later leaves reach the water's surface, the submerged leaves senesce and slough off, leaving the lower culms smooth.

As the leaves reach the water's surface, wax forms on the upper epidermis (Hawthorn and Stewart 1970). At this stage, internal tis-

Fig. 4. *(right top)* Microscopic view of a leaf of *Z. aquatica* var. *brevis* from which the lower (abaxial) epidermis has been removed and specially treated to show the diagnostic dumbbell-shaped costal silica bodies *(CS)* over a vein *(V)*. Long cells with wavy cell outlines alternate with papillae *(P)*, stomata *(S)*, and cork cells *(CC)*. Approximately 150×.
*(Right bottom)* A scanning electron microscope photograph of the abaxial surface of *Z. palustris* with stomata *(S)*, overarching papillae *(P)*, and extruded wax *(W)*. Exposed costal silica bodies *(CS)* are paired with cork cells *(CC)*. Approximately 400×.

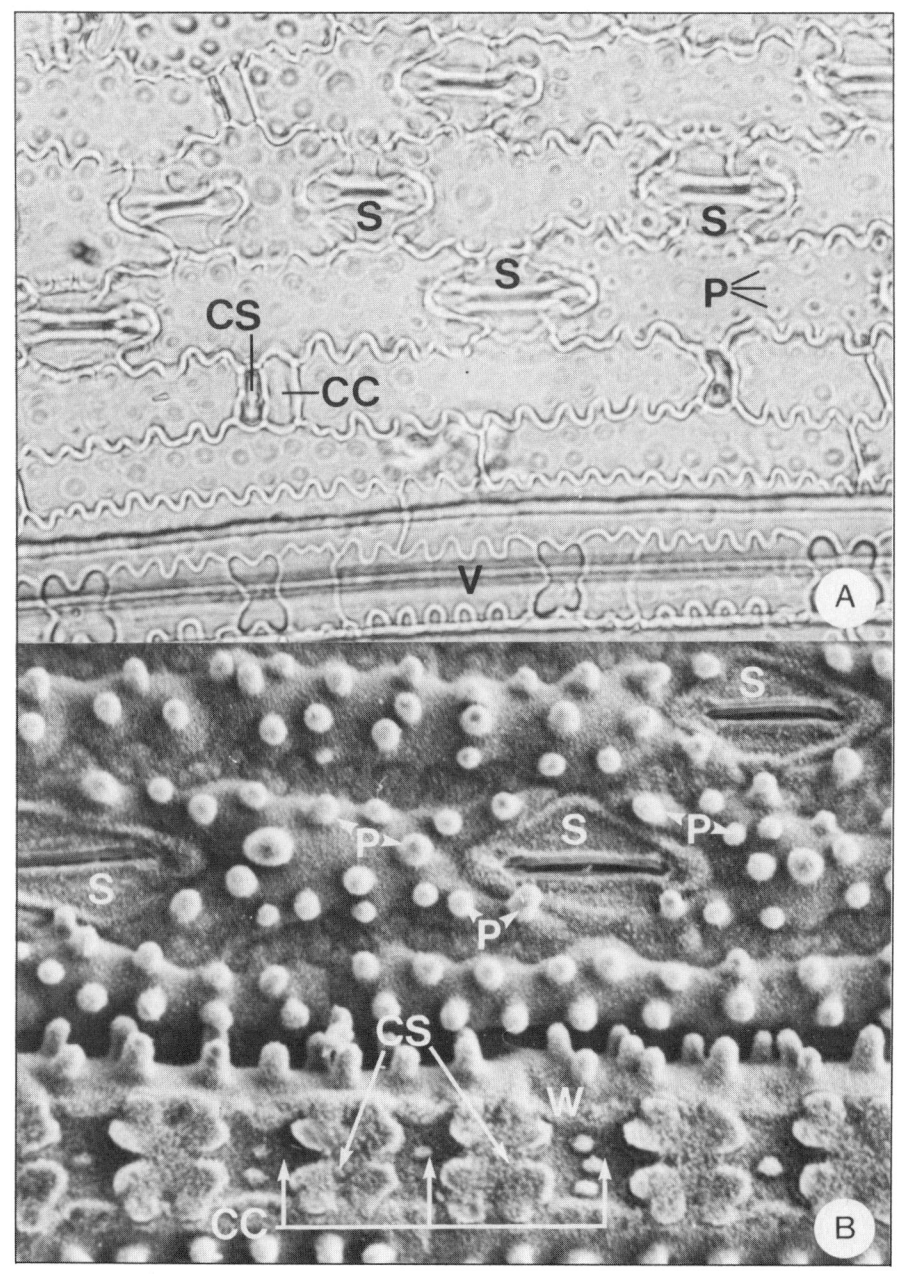

sue differentiation allows air to reach all parts of the plant. The long and ribbon-like floating leaves lie on the water's surface and are moved about by the waves and prevailing wind. There are usually two or three floating leaves per plant and they persist until the aerial leaves are well established.

When the aerial leaves emerge above the water's surface, their sheaths surround the culm. The blades are larger and more robust than those of the submerged or floating leaves. The upper and lower leaf blade surfaces are covered with epicuticular wax (Fig. 4). The leaf blades are commonly 30–70 cm long and 1–4 cm wide, depending on variety. The midrib is off center in the blade and more prominent on the lower surface. At the junction between the sheath and the blade is a scale-like appendage, or ligule, a membrane 3–11 mm long that is firm at the base but thinner and often frayed at the tip. The uppermost leaf on the stem is called the flag. It is shorter than the other aerial leaves and during its unfolding surrounds the developing inflorescence.

## Flowers

Separate male and female spikelets are borne on a terminal flowering panicle that is 40–55 cm long. Male spikelets (containing male flowers) are borne on lower flexible branches (Fig. 5). Above the male spikelets are stiff branches that bear the female spikelets (containing female flowers). This arrangement is the opposite positioning of the spikelets on corn where the terminal tassels are composed of many male spikelets. The segregation of the sexes promotes cross-fertilization because in the same panicle the female spikelets emerge and are mature 3 or 4 days before the male spikelets shed pollen. However, sometimes transition spikelets between the male and female ones have both stigmas and anthers and these could self-pollinate. Wind-pollination starts in late July and continues until the last of the tillers have flowered. Pollen viability is very sensitive to weather conditions. Recently, Terrell and Batra (1984) reported that bumble bees collect pollen from male spikelets but observed no visits by pollen-laden bees to female spikelets, and they concluded that these bees are not pollinators of wild rice.

The male spikelets vary in color from yellow to greenish pink or dark red. They hang down on thread-like stalks from the spreading panicle branches. The two outer scales (glumes) found in other grasses are reduced to a rudimentary ring of tissue in *Zizania*. Above this ring, two conspicuous scales, the lemma and palea, surround the six stamens and separate widely to expose the anthers. The stamens consist of long filaments that allow the anthers (pollen-bearing sacs) at the ends to hang outside the spikelets. Pollen is released about 2 hours after sunrise and by evening many male spikelets with empty anthers are shed. In the cultivated varieties the male spikelets remain on the plant, indicative of the non-shattering grain characteristic.

The female spikelets are borne erect from the panicle branches on short, firm stalks (Fig. 5). Each female spikelet has two conspicuous, interlocking scales, a lemma with a long, rough bristle (awn) at the tip and an unawned palea. As the spikelets push up above the leaf sheath, the palea kinks near the base of the spikelet, allowing the two white feathery stigmas to be exposed, one on either side. The number of female spikelets is variable depending on the variety and on environmental stresses to which the plant is exposed. In cultivated wild rice the female inflorescence produces up to 400 grains per panicle.

Wild rice harvesters report that not all female spikelets in a panicle produce mature grains. The amount of sterility varies and may be as high as 80% in southern wild rice. The causes of such sterility have not been studied, but they may be associated with both pollination and fertilization. Factors in the environment, such as wet, cold, or calm weather conditions, may interfere with the shedding of pollen or its transference to the stigmas.

## Grains

About 24 hours after the stigmas are pollinated, they wither and the ovary begins to enlarge with subsequent grain development. The grains, often called seeds, of wild rice are composed of an inner cary-

Fig. 5. *(next page)* Like corn, but unlike most other grasses, wild rice has separate male and female spikelets. *(a)* The male spikelets at the lower part of the panicle are just emerging from the leaf sheath and are not yet open, whereas those in the middle are in bloom and shedding their pollen. Toward the top, the withered spikelets have fallen off. Approximately 0.6×. *(b)* Female flowers at the top of the panicle in full bloom. Approximately 0.6×. *(c)* A male spikelet in full bloom, with six anthers exposed. Scale bar = 1 mm. *(d and e)* Two views of female spikelets in full bloom, with feathery stigmas exposed toward the base of the spikelet, through a gap where the palea has kinked outward. The stigmas shrivel soon after the pollen tubes penetrate them and fertilize the ovary, but they remain receptive for several days if they are not pollinated. When the stigmas begin to shrivel, the gap caused by the kinking outward of the palea closes. Approximately 2×.

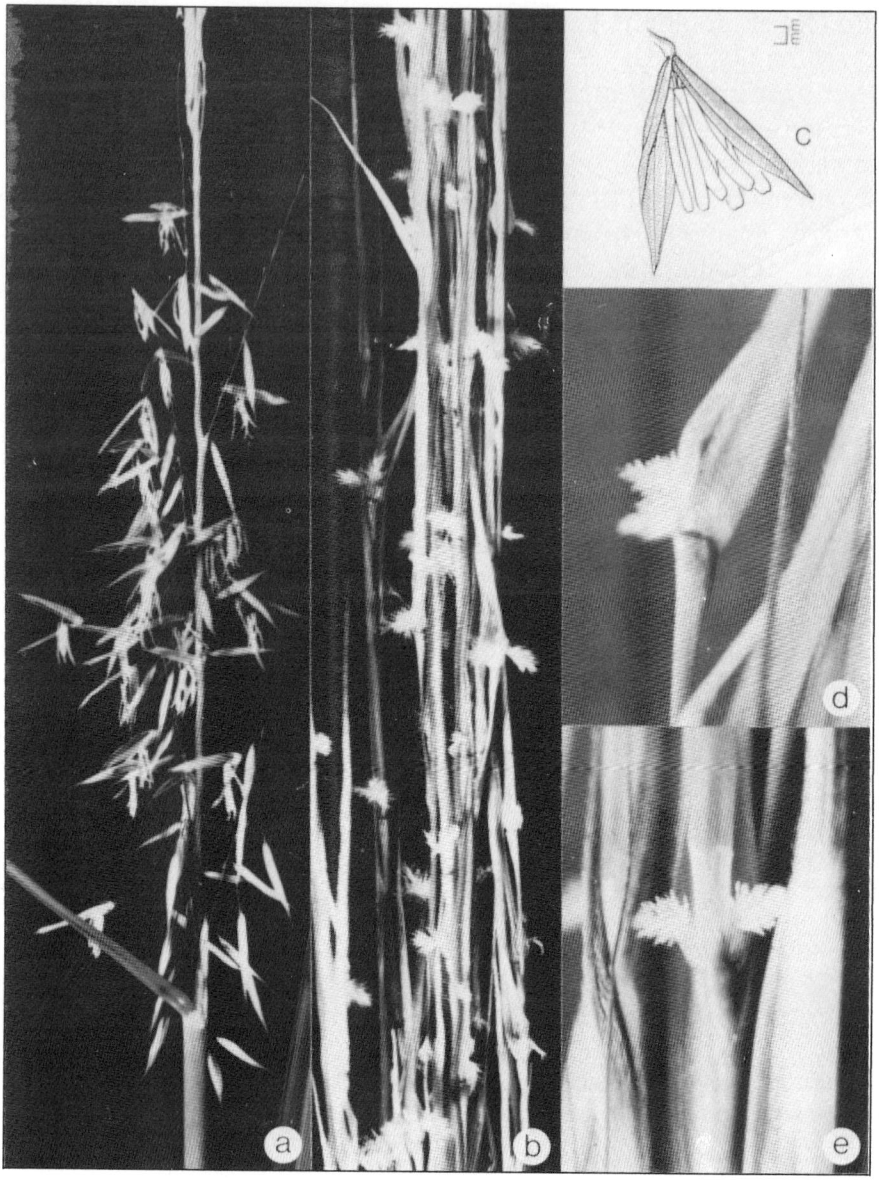

opsis that contains the embryo plant and food reserves, tightly surrounded by the lemma and the palea. The grains are shed about 4 weeks after fertilization (Elliott 1980). The grains change from light green with a firm-dough texture to olive brown and finally to dark brown/black with a hard kernel. Grain shedding occurs from 7 to 12 days after the firm-dough stage. Although many grains mature and shatter from the top of the panicle downward, there are plants in which maturation apparently takes place randomly throughout the panicle.

After leaving the parent plant, the grain falls through the water and plummets into the sediment, directed on its course by its rudder-like awn and the greater weight at the opposite end. Consequently it lodges close to the parent plant even in fast-flowing water.

A grain consists of an embryonic plant, or germ, that extends the whole length of the grain along the grooved side, and a reservoir of starchy endosperm. These structures are surrounded by a thin but tough and impermeable pericarp that contributes to seed dormancy. The grains of wild rice are longer than those of any other native grass and longer than most cereals in Canada. Each grain is tightly enclosed by the hull, which is made of the interlocking lemma and palea. The hull is not attached to the grain; hence the grain can be threshed clean following parching. However, it is difficult to remove the hull until it has been heated. The narrowness and relative brittleness of the dry grain make it susceptible to breakage when it is mechanically processed.

## Grain germination and viability

The percentage of wild rice germination can be high when grains are allowed to lie in their natural position at the bottom of a river over winter, or when stored in water at low temperatures. Grains incompletely matured on the stalk do not possess the same degree of germination. Grains fully ripened on the stalk have a definite dormancy and will not germinate for at least 3 months after ripening (Simpson 1966), even if temperature, moisture, and substrate are satisfactory for growth. They must pass through an afterripening period under freezing or near-freezing temperatures before the embryo breaks dormancy and develops into a new seedling plant.

There are conflicting opinions about the number of years wild rice grains remain dormant in the bottom of a body of water. Some observers maintain that plants can suddenly reappear in a pond after two, three, or several years of absence. Oelke et al. (1982) have recovered viable seeds from continuously flooded and summer-flooded soils 6 years after burial. Breeders have routinely fumigated their wild rice nurseries with methyl bromide as a means of eliminating this source of contamination (Elliott 1980).

When grain is stored dry, it rapidly loses its ability to germinate. In one of the earliest reported tests, Fyles (1920) observed that the rate of germination declined by 45% after 4 weeks and by a further 14% after 6 weeks. These observations, supported by the subsequent experiences of wild rice harvesters and farmers, indicate that wild rice grains intended for planting should not be allowed to dry out.

The development of wild rice cultivars for paddy fields in the United States and in Canada has required more than one generation of wild rice per year. A procedure to force germination was used by Elliott (1974). It involves removing the lemma and palea from dormant grains by forceps and breaking the impermeability of the pericarp by scraping. Germination proceeds when the scraped grains are submerged in water (Cardwell et al. 1978, Woods and Gutek 1974). Elliott (1974) reports that in a test of 43 lines using this scraping procedure, the success in plant establishment differed among the genetic lines. An average success rate of 30% suggested that factors other than pericarp presence are involved.

# II

# A Taxonomic Review of the Genus, Species, and Varieties

## Genus *Zizania*

The genus *Zizania* was named from a plant collected in Virginia by John Clayton in 1739. This specimen (Clayton No. 574) was sent to J. F. Gronovius in Leyden, Holland, for him to describe, which he did in 1743. The name chosen was from the Greek word *zizanion,* a weed of Mediterranean grain fields, thought to be the tares of the Scripture parable. In 1753, Linnaeus, the father of modern botanical and zoological nomenclature, provided the binomial *Zizania aquatica* for the Clayton specimen. Thus, it became the nomenclatural type for the genus. It is preserved in the Gronovian Herbarium at the British Museum of Natural History in London, England. The following are the main characters of this genus.

Plants annual, Asian *Z. latifolia* and North American *Z. texana* perennial, usually 0.25–3.5 m high but sometimes 4–5.5 m high; with a single stem (culm) or with tillers at the base, or rhizomes *(Z. latifolia),* the lower part of the culm sometimes decumbent and lying along the mud. Culm nodes are usually hairy, the internodes hollow and with stellate pith diaphragms present in the cavities (Fig. 3). The presence of these diaphragms allows rice straw to be distinguished from straw of other grasses or cattails. Aerial leaves 0.5–7.5 cm wide, with a ligule 3–25 mm long at base of blade and an open sheath surrounding the culm; aerial leaves have epidermal cells studded with low papillae and a thin, waxy cuticle. Silica deposits (Fig. 4), dumbbell-shaped in rows within epidermal cells, over vascular tissue, and scattered elsewhere in abaxial epidermis (Terrell and Wergin 1979, 1981).

The flowering stalk (inflorescence) is a terminal panicle 5–35 cm long, the lower branches are spreading with male spikelets that hang down, whereas the upper branches are usually upright with female spikelets stiffly erect. Widely spreading female branches, a condition known as "crowsfoot" inflorescence by plant breeders, are sometimes observed. Both male and female spikelets are borne on short, persistent stalks (pedicels). Shattering, or breaking, of these pedicels occurs just below a collar-like ridge of tissue that is thought to represent remnants of the vestigial glumes. Male spikelets usually drop as soon as pollen is shed; however, in cultivated varieties, and rarely, in native populations, male spikelets remain attached and are still conspicuous when the grain is ripe. They are (0.3–) 0.7–1.5 mm, have an outer, membranous bract (lemma) that is folded, with five veins, two of them at the margin, and a pointed tip or a short, straight awn. Inside the lemma are six anthers and a long bract (palea) that has three veins. Female spikelets are 2–20 mm long (without the awn). The lemmas have a straight awn usually about as long as the body of the lemma but often much longer, and 3–5 veins. Their margins curl inward, interlocking around the curled margins of the underlying palea and forming a zip-lock-like closure. At fertilization the palea is held at the apex by the enclosing margins of the lemma and kinks outwards opposite the top of the ovary, allowing the two fluffy white stigmas to be exposed (Fig. 4). After fertilization, the palea returns to its original position and remains tightly closed around the developing grain but is not attached to it. This interlocking lemma and palea structure is unusual in grasses and makes dehulling (removing the lemma and palea) more difficult in wild rice than in other cereal grasses. The parching process that dries and presumably shrinks the hulls releases the tight grip of the zip-lock.

The ovary, which is smooth and has two styles that are free to their bases, matures to form a cylindrical caryopsis 1–2 cm long. Inside, an embryo extends the full length of the grain. The base chromosome number for the genus is $x = 15$. All the North American taxa have the same chromosome number $2n = 30$ (Brown 1950, Dore 1969). The chromosome number of the Asiatic *Z. latifolia* is $2n = 34$ and this species produces sterile hybrids when crossed with *Z. palustris,* one of the annual North American species (Bondar 1958).

## Taxonomic categories within the genus *Zizania*

Two species of wild rice are perennial. One is native to Asia, occurring from Manchuria to northeastern India; the other is native and very localized along the San Marcos river in Texas. The annual varieties of wild rice, which are native to North America, can conveniently be divided into four categories, but there have been differences of opinion as to whether these categories should be treated as one or two species. The categories recognized by Dore (1969) and used here, as well as the names suggested for them by other workers, are summarized in Table 1.

Table 1. Categories within the genus *Zizania*

| Categories used by Dore (1969) | Categories used by Fassett (1924), Fernald (1950), and Gleason and Cronquist (1963) |
|---|---|
| **Annuals Found in Canada** | |
| 1a. *Z. aquatica* var. *aquatica* L. **southern wild rice** | *Z. aquatica* L. *Z. palustris* L. |
| 1b. *Z. aquatica* var. *brevis* Fassett **estuarine wild rice** | *Z. aquatica* var. *brevis* Fassett |
| 2a. *Z. palustris* var. *palustris* L. **northern wild rice** | *Z. aquatica* var. *angustifolia* A. S. Hitchc. |
| 2b. *Z. palustris* var. *interior* (Fassett) Dore **interior wild rice** | *Z. aquatica* var. *interior* Fassett |
| **Perennials Found Elsewhere** | |
| 3. *Z. texana* A. S. Hitchc. **Texas wild rice** | |
| 4. *Z. latifolia* Turcz. **Manchurian water-rice** | |

Dore (1969) based his recognition of two species of annual wild rice *(Z. aquatica* and *Z. palustris)* on his experience and his interpretation of the original concepts of Linnaeus who proposed the two species names at different times but without cross reference. Unfortunately some later European and American authorities have not agreed on the application of these names or even on their separate validity. In North America, when A. S. Hitchcock reviewed the genus in the seventh edition of *Gray's Manual of Botany* (1908), he accepted Linnaeus's names as representing two distinct species but deliberately applied the names in reverse order, recognizing *Z. palustris* L. (= *Z. aquatica* of auth., not L.). This mistake, appearing in such an important manual, accounts for the confusion that continues to arise whenever reference is made to records or specimens accumulated between 1908 and 1924. Chambliss's work, published by the U.S. Department of Agriculture in 1922, followed Hitchcock's treatment, but the eighth edition of *Gray's Manual of Botany* (Fernald 1950) reversed the use of the names, recognizing *Z. aquatica* (Typical) (= *Z. palustris* of 7th ed., not L.).

Wiegand and Eames (1925) used the two names correctly, but their treatment was eclipsed by the critical monograph of Fassett (1924) in which he recognized one species, *Z. aquatica,* and four varieties. Fassett (1927) considered it necessary to reiterate his concept and to explain that, in his opinion, the variety *interior* in its morphology completely bridged the gap between the two proposed species. Fassett's treatment was followed by the major works of Fernald (1950) and Gleason and Cronquist (1963).

The question of the taxonomic level appropriate for the categories of annual wild rice has persisted. Initial results from isoenzyme studies among the taxa found in Canada strongly favor the recognition of two species (Warwick and Aiken 1986). Dore (1969) suggested that distinct species status may be appropriate for the variety *brevis,* because its short awn, small grain, and short height are characteristic and perpetuated under cultivation. The unique tidal habitat of var. *brevis* imposes distinct and intense selection pressures not affecting other varieties. However, a scanning electron microscope study of epidermal features of this taxon showed only minor ways in which it differs from *Z. aquatica* (Darbyshire and Aiken 1986) and no isoenzyme distinctions were found (Warwick and Aiken 1986).

Geographic separation is the chief factor that keeps varieties within the two species distinct. When they occur at the same place, intermediate individuals are usually found. Plants that look intermediate between varieties *palustris* and *interior* occur often, but at least some of these result because northern wild rice grown in nutrient-rich conditions produces large plants that look like interior wild rice plants grown in nutrient-poor conditions. Isoenzyme distinctions were found between varieties *interior* and *palustris* (Warwick and Aiken 1986).

There are some unusual plants localized in Bear Brook, near Ottawa, with many of the general features of the southern variety but with short awns. They were first pointed out by Fassett (1924) and named variety *subbrevis* by Boivin (1967). However, *Zizania aquatica* in the southern United States has spikelets with very variable awn lengths, not uncommonly only 0.2–0.6 cm long (E. E. Terrell, pers. commun., 1984). Dore and McNeill (1980) speculated that Bear Brook on the South Nation drainage system was influenced by estuarine waters of the retreating Champlain Sea; the variety *subbrevis* may represent the descendants of plants adapted to tidal conditions. They also observed that the type specimen of var. *subbrevis* happens to be considerably dwarfed in stature because it represents a plant browsed on during its early growth, and they refer to several other collections from the South Nation drainage system that reveal an almost complete gradation from varieties *aquatica* through *subbrevis* to *brevis*. Isoenzyme tests on collections from Bear Brook gave results identical with those obtained for *Z. aquatica* from other sites (Warwick and Aiken 1986). Thus, var. *subbrevis* appears to represent a portion of an ecocline within *Z. aquatica*, another portion of which was recorded for a population of particularly large plants from the Wading River, New Jersey (Ferren and Good 1977).

Within varieties, individual plants have inherent morphological variations of a minor nature. Their variations include differences in color of the plant, height and number of culms, length of the spikelet awns, size and shape of the grain, number of grains to a branch, time of ripening, and how readily mature panicles shatter. Selection by man for agronomically desirable qualities in wild rice has resulted in cultivars that are sometimes called varieties commercially but should not be confused with taxonomic varieties.

## Key to the annual varieties of wild rice

1. Female inflorescence branches usually spreading at maturity; lemmas of female spikelets delicate, thin, papery, dull-opaque, scabrous over the whole surface and slender vein ribs; paleas scabrous; aborted spikelets[1] appearing thread-like (Fig. 6), less than 1 mm thick. Male spikelets usually less than 1.5 mm wide before anthesis.
    2. Plants usually (0.8–)1–3(–5.0) m high; leaves 10–80 mm wide; ligules (6–)10–20(–25) mm long; body of mature female lemma 10–20 mm long, usually minutely roughened; awn usually 10–75 mm long (Fig. 7).
    ...... 1a. *Z. aquatica* var. *aquatica*
    2. Plants usually 0.25–1 m high; leaves 3–12 mm wide; ligules about 3 mm long; body of mature female lemma 5–10 mm long, smooth or roughened; awn 1–8 mm long; freshwater tidal flats of the lower St. Lawrence River.
    ........ 1b. *Z. aquatica* var. *brevis*.

1. Female inflorescence branches usually remaining erect and adpressed at maturity; lemmas of female spikelets firm, tough, straw-like, lustrous, scabrous on the margins at the apex, along the awn, and sometimes along the prominent corrugated vein ribs, elsewhere glabrous; paleas glabrous; aborted spikelets[1] not thread-like (Fig. 6), 1.5–2 mm thick. Male spikelets usually 1.5–2 mm wide before anthesis.
    3. Plants 0.7–1.5 m high; leaves 4–12(–15) mm wide; ligules 3–5(–10) mm long; lower female branches with 2–6 spikelets; lower or middle male branches with 1–15 spikelets (Fig. 8).
    ...... 2a. *Z. palustris* var. *palustris*
    3. Plants 0.9–3 m high; leaves 10–30(–40) mm wide; ligules 10–15 mm long; lower female branches with 10–30 spikelets; lower or middle male branches with (20–)30–60 spikelets.
    ....... 2b. *Z. palustris* var. *interior*

---

[1] There is an unfortunate mistake in the keys presented in Dore and McNeill (1980) with the word "male" appearing where the word "sterile" was intended (W. G. Dore, pers. commun. 1985).

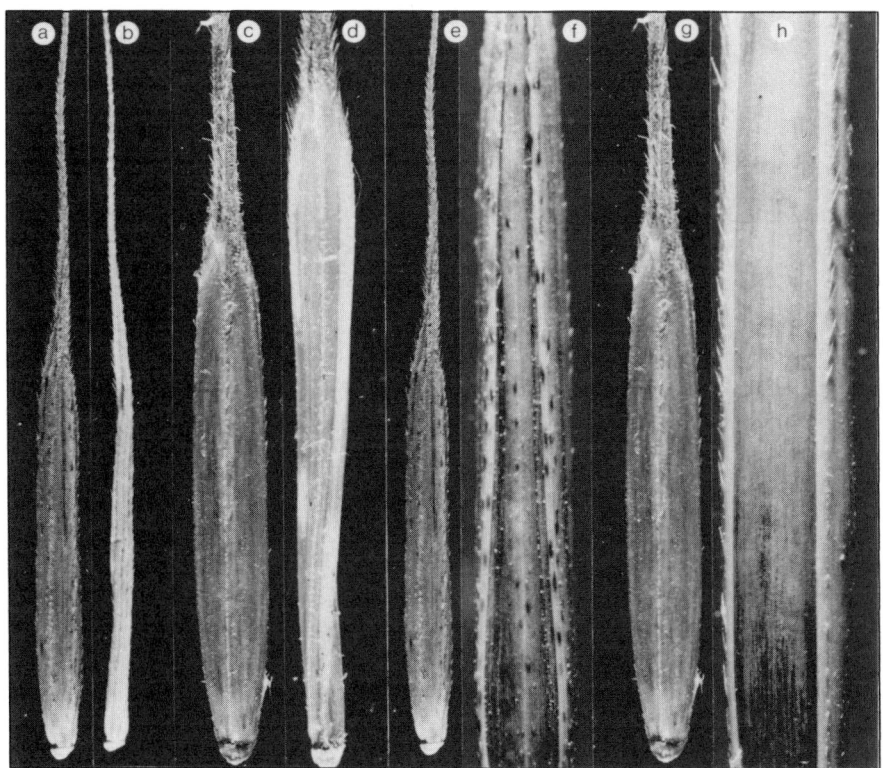

Fig. 6. A comparison of sterile and fertile spikelets of Z. aquatica (a, b, e, f) and Z. palustris (c, d, g, h). (a and c) Fertile spikelets. Approximately 3.5×. (b) The sterile spikelet of Z. aquatica is rounded and thread-like. (d) The sterile spikelet of Z. palustris is flattened and 1–2 mm wide. (e) Dorsal view of Z. aquatica spikelet showing scabrous prickles along the veins from the apex to the base and scattered prickles on the lemma surface. Approximately 8×. (f) Ventral view of spikelet of Z. aquatica showing scabrous prickles on the margins of the lemma and on the palea surface. Approximately 3.5×. (g) Dorsal view of a spikelet of Z. palustris showing scabrous prickles extending along the veins near the lemma apex, but elsewhere the lemma is glabrous. Approximately 3.5×. (h) Ventral view of spikelet of Z. palustris showing scabrous prickles along the veins at the margins of the lemma and a glabrous palea surface.

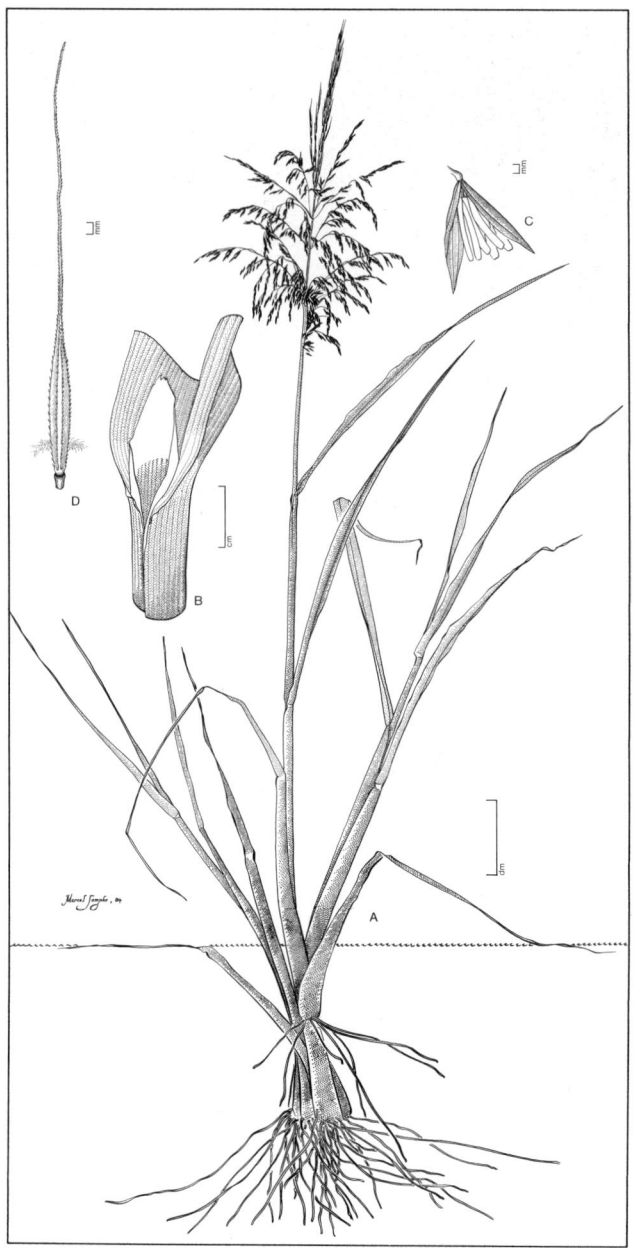

Fig. 7. *Zizania aquatica.* (A) Flowering plant, approximately 2 m high. (B) Junction between leaf sheath and blade showing ligule. Scale bar = 1 cm. (C) Pendulous male spikelet at anthesis showing six anthers. (D) Erect female spikelet with long awn. Scale bar = 1mm.

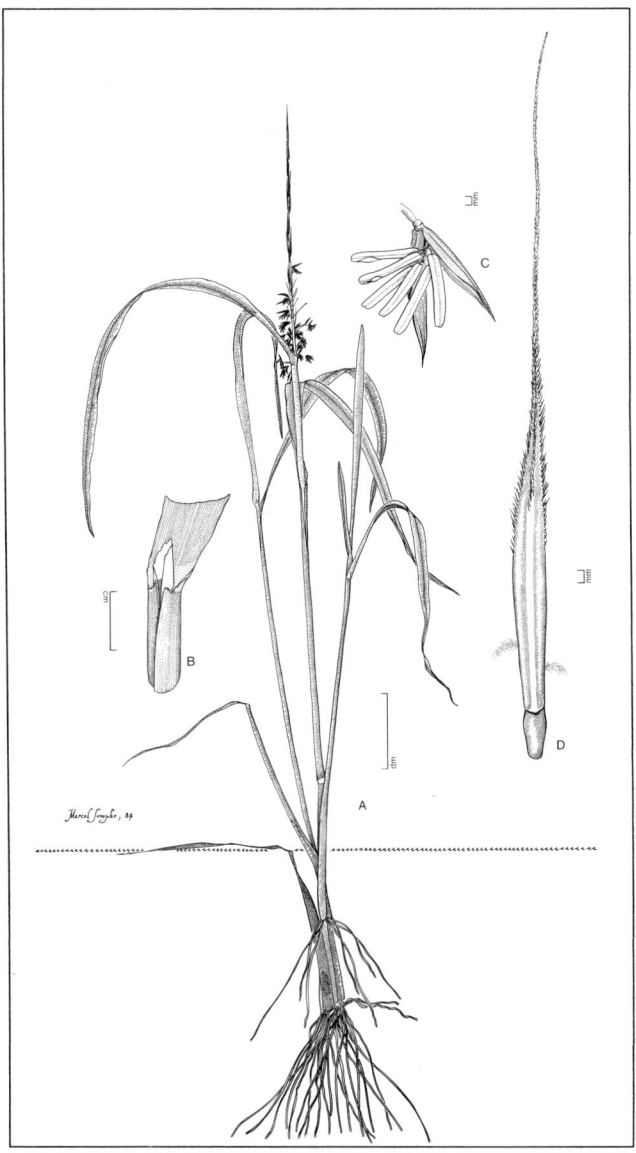

Fig. 8. *Zizania palustris*. *(A)* A flowering plant approximately 1.5 m high. *(B)* The junction between leaf sheath and blade showing ligule. Scale bar = 1 cm. *(C)* Pendulous male spikelet at anthesis, showing six anthers. Scale bar = 1 mm. *(D)* Erect female spikelet with long awn. Scale bar = 1 mm.

## *Zizania aquatica*

1. *Zizania aquatica* L. Sp. Pl. ii. 991. 1753.

**Type:** United States, Virginia. J. Clayton 574 before 1739. (Gronovian Herbarium, at British Museum.)

Annual, 0.25–3.0 (rarely to 5.5) m high; hull of grain thin, papery, dull, and minutely roughened on the surface; aborted spikelets shrinking and becoming thread-like, less than 1 mm thick. Two varieties are recognized.

1a. *Zizania aquatica* var. *aquatica* L. SOUTHERN WILD RICE, SOUTHERN ZIZANIA. This is the typical variety of the species (Fig. 7).

Plants usually (0.8–)1–3(–5.0) m high; culms often 2–3 cm thick at base; leaf blades 1–8 cm wide, pale green, usually drooping outward at the top; panicles large, usually 25–80 cm long, many-branched, with numerous spikelets; female spikelets often sterile; awn usually 1–7.5 cm long.

Southern wild rice is present in Canada along the shores of the Great Lakes and along muddy shores of streams and ditches in southern Ontario and Quebec. It occurs throughout Florida, Louisiana, and the eastern seaboard states. Because the grain is short and thin there is little commercial interest in it, and the places where it is now found in Canada are mostly sites of natural distribution.

Dore (1969) documented the limited and disjunct distribution of this southern variety in Canada and concluded that because of its lack of aggressiveness, it is unlikely to extend northward under present climatic conditions. The map (Fig. 9) is essentially the same as that published by Dore.

1b. *Zizania aquatica* var. *brevis* Fassett. Rhodora 26:157. 1924.

**Type:** Canada, Quebec, St. Lawrence River, rocky tidal flats, Levis. 9 Aug. 1923, H. K. Svenson and N. C. Fassett, no. 853 (Gray Herbarium). ESTUARINE WILD RICE, ESTUARINE ZIZANIA.

Plants 0.25–1 m high; culms slender, flexible; leaf blades 3–8 (–12) mm wide, dull green; panicles 10–25 cm long, sparsely branched, few spikelets; female spikelets with an awn usually less than 1 cm long.

Estuarine wild rice grows in the extensive tidal flats of the St. Lawrence River estuary where, although tides occur twice a day, the water is fresh rather than salty. During progressive inundation by the tide, the plants offer little resistance. Their stems lean forward

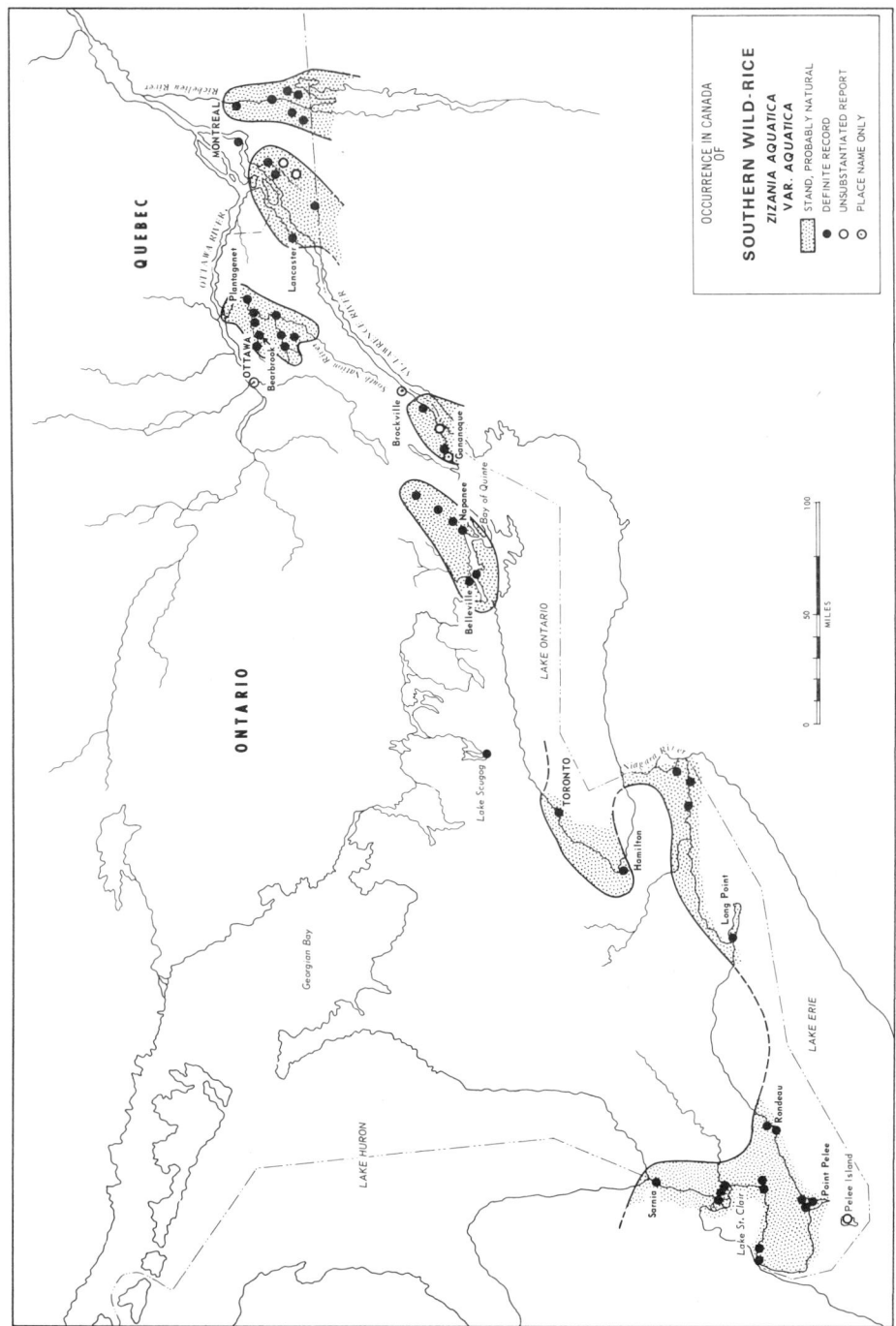

Fig. 9. Map of the distribution of *Z. aquatica*.

with the water currents and eventually the plants become completely submerged. The performance is repeated as the tide flows out. Often drifting debris and freshwater algae are caught in the panicle and a veneer of mud is left on the whole plant. The plants cannot tolerate very salty water and this intolerance defines the lower limit of their range: about 80 km downstream from Quebec City on the south shore, at Trois-Saumons, and about 30 km downstream on the north shore, at Sainte-Anne-de-Beaupré and at the northeastern tip of the Île d'Orléans. The variety occurs almost continuously upstream for about 80 km from Quebec City to Grondines (Fig. 10).

There are morphological differences in the plants, apparently in response to the saline gradient along the estuary; thus, where the salt concentrations are higher, the green stems and leaves have a somewhat succulent or rubbery texture not found in plants upstream. These plants also grow remarkably tall for the variety, and develop long, stiff, and spreading panicle branches. The anatomical basis for these intravarietal differences has not been investigated, but the differences are presumed to be largely environmental rather than inherent.

Plants reach their optimal development and bloom earlier on the shoreward portion of the flats, where, when the tide is in, they are submerged for a shorter time. A few are inundated only at times of high spring tides. Probably the water temperature and the varying interval of submergence affect development time. Consequently, ripening of grain takes place over an extended period. This, and the small size of the grain, are the reasons estuarine wild rice has never been harvested commercially. Nevertheless, the plant provides a valuable crop of food for migratory birds and other wild fowl that flock to the tidal flats each year. Several commercially desirable morphological features are expressed in this variety and it may well prove to be a useful source of genes in future crop development.

Estuarine wild rice is the only variety of *Z. aquatica* that grows in the tidal flats of the St. Lawrence River estuary, and it is not known from other tidal estuaries along the Atlantic coast. A logical conclusion is that it has evolved in the present or former estuary of the St. Lawrence and is endemic to that region.

## *Zizania palustris*

2. *Zizania palustris* L. Mant. 295. 1771.

**Type:** The type specimen of *Zizania palustris* L. is preserved in the

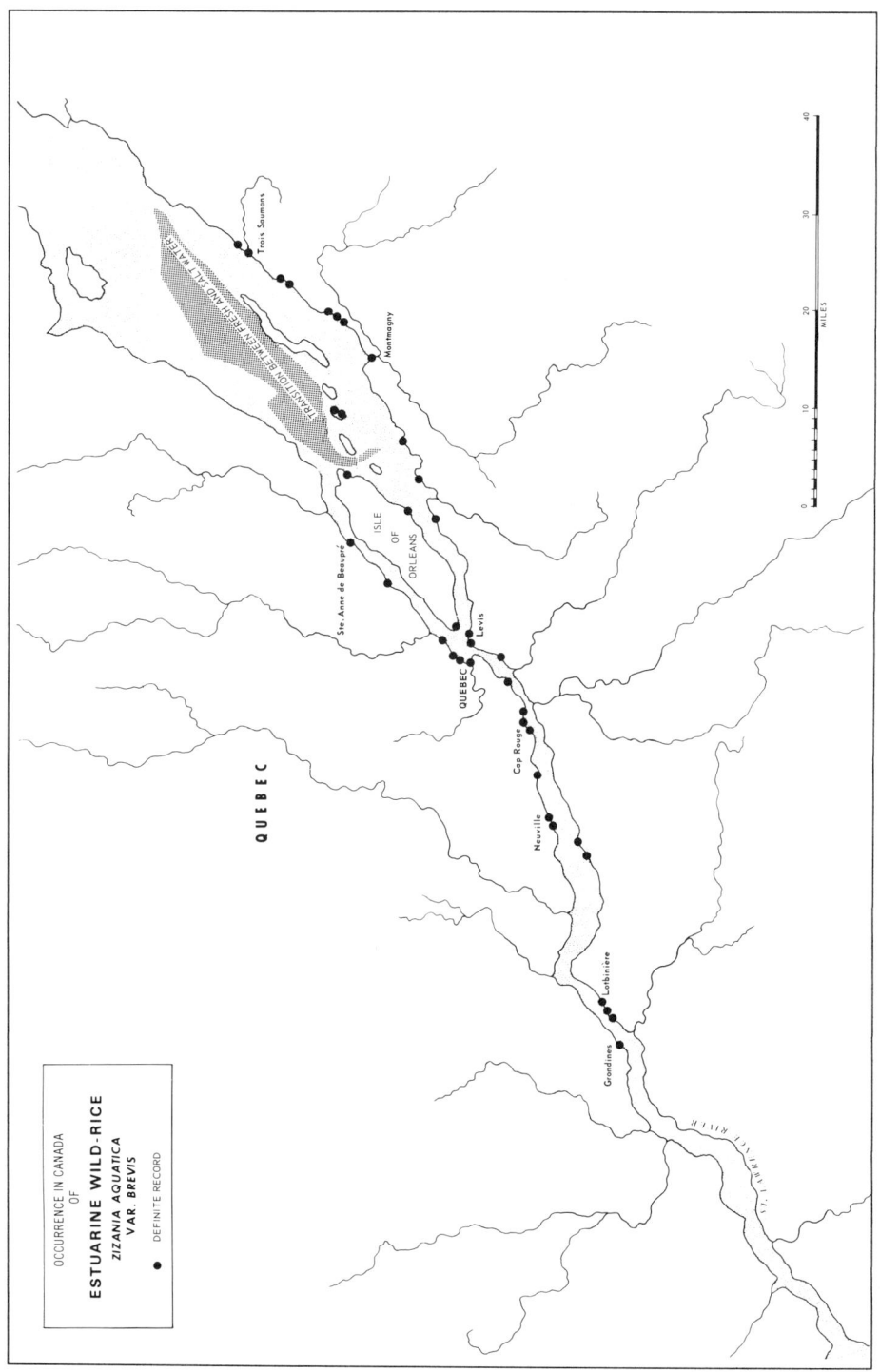

Fig. 10. Map of the distribution of Z. *aquatica* var. *brevis*.

British Museum of Natural History. On the herbarium sheet Linnaeus wrote only *"Zizania"* at the left of the base of the stalk and "H U" at the right, which meant the plant was grown in his garden, *Hortus Upsalensis,* at Uppsala, Sweden. From other sources we know that the grain was collected near Montreal by his student, Pehr Kalm, in 1749. The other annotation faintly showing on the sheet was made later by the authority J. E. Smith: *"palustris vix ob aquatica diversa"* (*palustris,* scarcely different from *aquatica*), an opinion that led later students to doubt the specific distinctness Linnaeus wished to designate.

Annual, 0.5–3 m high; hull of grain firm and leathery, shiny and smooth on the surface but with scabrous prickles along the veins. Two varieties are recognized.

2a. *Zizania palustris* var. *palustris* L. (*Z. aquatica* var. *angustifolia* A. S. Hitchc.) Rhodora 8:210. 1906.

**Type:** United States, Maine, Belgrade. Aug. 1895, F. L. Scribner. (Gray Herbarium.) NORTHERN WILD RICE, NORTHERN ZIZANIA (Fig. 8).

Stems slender, 0.5–1.5 m high; leaf blades 4–12(–15) mm; panicles slender and few-flowered; male spikelets usually less than 15 on a branch, female spikelets 2–6 on a branch.

Northern wild rice grows extensively along many rivers or their lakelike expansions from the southern edge of the Precambrian area in Alberta and Saskatchewan (range extended by planting) through Manitoba and northwestern Ontario. It is widespread in southern Canada where considerable stands are found in waters of the Trent Canal System and rivers flowing into the Ottawa from the north and west. It occurs as far east as Nova Scotia and extends southward into the northern United States (Fig. 11).

Northern wild rice is the most common and commercially important variety in Canada. This variety is popular among private and governmental sectors for plantings in rivers and lakes. As a consequence of this planting the range of this variety has been extended and there is an admixture of tall broad-leaved types that intergrade with the next variety.

2b. *Z. palustris* var. *interior* (Fassett) Dore. Can. Dep. Agric. Publ. 1393. 1969.

**Type:** United States, Iowa, Armstrong. 27 Aug. 1897, L. H. Pammel and R. J. Cratty, no. 764 (Gray Herbarium). INTERIOR WILD RICE, INTERIOR ZIZANIA.

# A TAXONOMIC REVIEW OF THE GENUS, SPECIES, AND VARIETIES 35

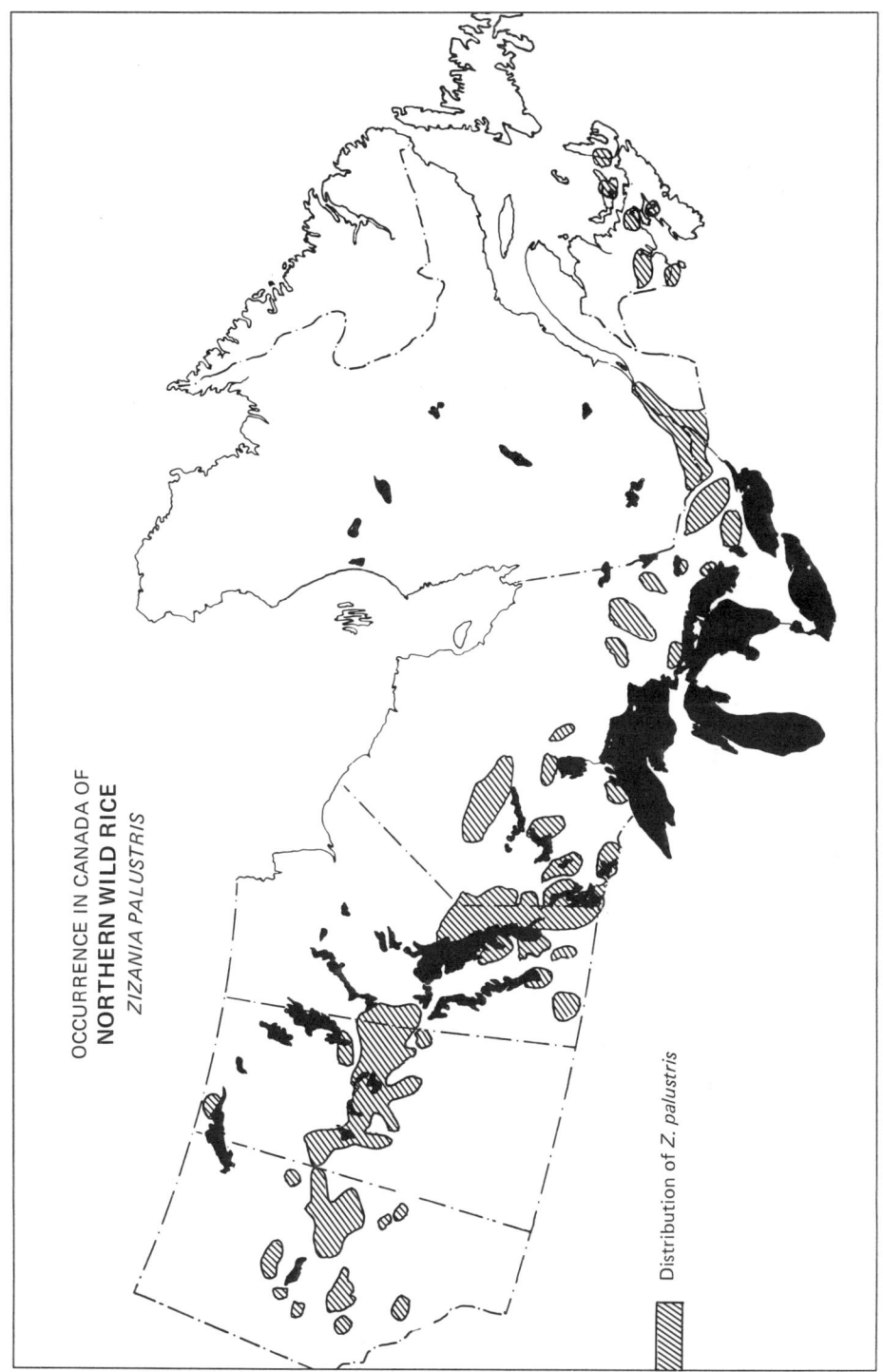

Fig. 11. Map of the distribution of *Z. palustris*.

Stems usually 0.9–3 m high; leaf blades 10–40 mm wide; panicles spreading, branches generally divergent and many-flowered; male spikelets (20–)30–60 on a branch, female spikelets 10–30 on a branch. Interior wild rice grows on muddy shores and in water up to 30 cm deep mainly in the North Central United States. In Canada, natural stands occur only in the southernmost portion of Manitoba. Scattered stands occur at various points where it has been planted, from Manitoba and adjoining Ontario, to New Brunswick where it is abundant in the lower St. John River. In natural habitats it grows where the water retreats in the summer leaving flats of muddy alluvium. It forms spikelets and ripens earlier than the northern variety. The grains are somewhat shorter than those of the northern variety, but they are produced in greater abundance.

The two varieties are closely related and often difficult to distinguish morphologically, especially when plants are not fully developed. Variety *interior* grain is harvested throughout Alberta, Saskatchewan, Manitoba, and northern Ontario and often sold for new plantings. Thus, the northern and interior varieties come to grow together and many apparently intermediate forms are found making it difficult to separate the two varieties. However, isoenzyme research indicated ways in which the two varieties can be distinguished (Warwick and Aiken 1986).

## *Zizania texana*

3. *Z. texana* A. S. Hitchc. J. Wash. Acad. Sci. 23:454. 1933.

**Type:** United States, Texas, San Marcos. April 1932, W. A. Silveus, no. 518. (U.S. National Herbarium, no. 1, 174, 537). TEXAS WILD RICE, TEXAS ZIZANIA.

Culms many, decumbent, geniculate, rooting at the lower nodes, forming stolons 1–3 m long (sometimes to 5 m), immersed culms and leaves long-streaming in river currents with only the inflorescence emergent, or in slow currents with upper culms and leaves emergent. Basal sheaths yellowish; ligules usually dark basally, whitish distally, 4–12 mm long, 2–5 mm wide. Submersed leaves to about 1 m long, 7–13(–25 mm) wide. Panicles 16–31 cm long, 1–10 cm wide. Male branches erect or somewhat spreading, to 10 cm long, pedicels 1–9 mm long, expanding to about 0.3 mm in diameter at base of the male spikelets, these 6.5–11.0 mm long, 1.2–2.0 mm wide. Female branches to 7 cm long, pedicels 1–7(–13) mm long, expanding to 0.5–0.9 mm in diameter at base of female spikelets, these 9.0–12.5 mm

Fig. 12. Manchurian wild rice, *Zizania latifolia*, grown by Dr. W. G. Dore in a muddy paddy at Ottawa. This broad-leaved Asiatic species is a perennial, but when grown outside or indoors in the greenhouse it never produced flowers. The clumps were grown from small rhizome pieces set out the previous fall and they produced stems over 1.5 m high.

long, and 1.2–1.8 mm wide, leathery, light brown, greenish, or with the basal half greenish and the distal half whitish, the surface with scattered prickle hairs and the apex tapering to a terminal awn 9–35 mm long. Aborted spikelets 7.5–12.0 mm long, 0.7–1.2(–1.5) mm wide. (Description based on Terrell et al. 1978.).

Known only from Texas, Hays County, in the vicinity of the San Marcos River, near its source, where it grows in the cool, fast-flowing spring-fed water and is considered to be endemic. The stands are reported to be decreasing in size because of increasing pollution in the area (Emery 1977).

There is reason to believe that its perennial nature is due to the constant year-round temperature of the artesian waters. The status of *Z. texana* as a species distinct from *Z. aquatica* was questioned by Dore (1969), but the epidermal features studied by Terrell and Wergin (1981) suggest that it is a distinct species.

## *Zizania latifolia*

4. *Z. latifolia* (Griseb.) Stapf. Kew Bull. 385. 1909. Turczaninow Bull. Soc. Nat. Moscow. 11(1):105, nom. nud. MANCHURIAN WATER-RICE, MANCHURIAN ZIZANIA (Fig. 12).

Plants perennial, spreading by coarse subterranean runners; male and female spikelets borne on stems having a row of microscopic hairs at the tip and a corresponding crown of hairs present at the base of the hull of the floret; hull of grain thin, papery, dull, and rough, as in *Z. aquatica*. In contrast with other species, there are microhairs scattered over the body of the female lemmas (Terrell and Wergin 1981).

It is a native grass of Manchuria, Korea, Japan, Burma, and northeastern India. In parts of the Orient, some stands of *Z. latifolia* are sterile because of infection by a systemic fungus *(Ustilago esculenta* Hennings; Terrell and Batra 1982). The swollen shoots of such plants are highly prized as a vegetable. This species also has potential as a forage grass, and it has been planted for grazing purposes in other countries of Asia and Europe. Flowering plants grown at the Royal Botanic Gardens at Kew provided the material from which Dr. Stapf prepared the first detailed description and compared it with the North American *Z. aquatica*. However, plants of this species grown for several years in a botanical garden and greenhouse in Ottawa, and plants established in a greenhouse at the University of Minnesota, never flowered. At the Patuxent Reserve, near Washington, where the species was first introduced for testing in the 1920s, the plants always flowered, but in some seasons too late to ripen grain (Chambliss 1922).

# III

# Wild Rice Habitat

Wild rice grows best in shallow, clear lakes or rivers with a soft, organic bottom into which a paddle can easily penetrate, where there is an absence of plant competitors. However, wild rice does grow in a wider range of aquatic environments. Studies on the habitat requirements of wild rice have been few and the following information is based primarily on research done in Minnesota and northwestern Ontario.

## Water depth

Water depth is the most important factor influencing a wild rice crop. The ideal depth during the plant's life cycle is about 0.3–0.6 m. If the water is either too deep or too shallow, or if sudden fluctuations in depth greater than 25 cm occur, production is severely affected. Moyle (1944) determined that an increase in depth of 0.3 m from preceding years could almost eliminate the wild rice crop on certain lakes in Minnesota. Similarly, on Lake of the Woods in northwestern Ontario, an increase of mean summer depths of 0.6 m from 1973 to 1974 decreased the harvest by over 400 000 kg (Lee 1975a). The problem with deep water is that it does not allow sufficient light penetration for normal photosynthesis to occur. Under such conditions seedlings may die or may have limited growth at their emergence stage. This is particularly apparent in darkly stained lakes, where wild rice grows only at the littoral zone (Fig. 13). Tillering, the production of many stems from one plant, seems

Fig. 13. A lake with darkly stained water in which wild rice grows only in the littoral regions.

to be light induced and is reduced in deeper waters (Lee 1979).

Although water that is too deep is more critical, very shallow waters can affect wild rice production. Weber and Simpson (1967) found that if the depth fell below 10.4 cm during the submersed or floating-leaf stage, the grain yield and amount of dry matter produced were reduced. Thomas and Stewart (1969) reported that a water depth of at least 8 cm was required for normal development.

Sudden fluctuations in water depth are also detrimental to rice production. If the water depth is increased during the floating-leaf stage while the roots are still small, the wild rice plants may be uprooted as the buoyancy in the leaves lifts the roots out of the soft sediments (Chambliss 1940). At the floating-leaf stage, a cuticle forms on the leaves, and the plant is no longer able to obtain dissolved gases from the water over the entire unwaxed leaf surface as it did during the submersed-leaf stage. Now it obtains the gases through small pores (stomates) that develop in the upper waxy surface of the floating leaves (Hawthorn and Stewart 1970). If the leaves are flooded, the stomates close and the plant is unable to take up sufficient amounts of gases through these leaves. During the aerial-leaf stage, if water depth is increased too rapidly, the lower branches of the inflorescence with the male spikelets may be under water and wind pollination cannot occur.

## Water chemistry

Wild rice grows in a wide range of water types. In a survey in northwestern Ontario and northeastern Minnesota, the water quality of lakes with wild rice was found to be of two types (Lee 1979). The first had an alkalinity of about 40 mg/L and pH values of about 6.9. This corresponds to Moyle's (1944) description of soft-water flora in Minnesota, which he described as those species living in waters with a total alkalinity less than 40 mg/L, a pH between 6.8 and 7.5, and a sulfate concentration less than 5 mg/L. Moyle included wild rice in this group. The second type of wild rice lake found in Lee's study had an alkalinity of about 80 mg/L and pH values of 7.4. These values were somewhat lower than Moyle's hard-water flora, which he described as occurring in lakes with alkalinities of 90–150 mg/L, a pH of 8.0–8.8, and a sulfate concentration of 5–40 mg/L. Moyle (1945) suggested that sulfate was particularly important in influencing wild rice distribution, with few stands in Minnesota occurring in waters having a sulfate concentration greater than 40 mg/L and no large stands occurring in surface water having sulfate concentrations greater than 10 mg/L.

Evidence exists that wild rice absorbs most of its required nutrients from the soil, which suggests that water chemistry is only indirectly important. Tissue analyses showed that the seasonal trends of inorganic nutrients in wild rice roots were similar to those in the stems and leaves (Lee and Stewart 1983). In controlled greenhouse studies, sulfate concentrations in the water overlying a silica sand substrate, described by Moyle as adversely affecting wild rice, were shown to have no effect on wild rice growth (Lee and Stewart, Unpubl. Rep. 1978). This indicates that the water chemical values suggested by Moyle are important only to the extent that they correlate to the nutrients levels in the underlying soil. In some cases, as in the above study, they clearly do not. In another study (Davis 1979) an average sulfate concentration of 170 was found in the waters of very productive wild rice paddies along the Clearwater River in Minnesota. Another problem from correlating water chemistry to soil chemistry is the tendency for the nutrients in the water to fluctuate greatly in wild rice stands throughout the summer (Lee and Stewart 1981). Such fluctuations in the chemistry of the water, caused simply by time, negate the possibility of relating the concentrations of nutrients in the water to those in the soil.

The need for an adequate oxygen supply may explain the importance of a good flow of water through a wild rice stand, often noted by early researchers (Dore 1969). The gases dissolved in water are important to the submerged stages of wild rice. In highly eutrophic conditions, reduced carbon dioxide has been shown to limit production of wild rice (Lee and Stewart 1981).

## Type of soil

Wild rice is known to grow on a wide range of soil types, but it usually grows best on soft alluvial organic soil (Brown and Scofield 1903, Fyles 1920, Lloyd 1939, Chambliss 1940, Moyle 1944). The roots of the rice plants easily penetrate this substrate (Chambliss 1940), and because this type of soil erodes easily, competing perennial species must often be eliminated (Dore 1969). One perennial species, however, actually enhances the growth of wild rice. Northwestern Ontario lakes over mineral soils only become optimal for wild rice after the lakes have been colonized by the underwater plant *Potamogeton robbinsii* (Fig. 14), which adds organic matter to the sediment (Lee 1983*a*).

The submersed condition of soils in which wild rice grows results in characteristics that are different from terrestrial soils. Oxygen is usually limiting, and this causes many of the nutrients required for wild rice growth to occur in their reduced state. Nitrogen is present as ammonium; sulfur as sulfide; and the metals, iron (Fe), manganese (Mn), zinc (Zn), and copper (Cu), in their lower valence states. Changes in the chemical nature of these elements, in turn, affect the availability of other required nutrients. In flooded organic soils, the reduced form of iron does not bind phosphorus and causes phosphorus to be released and become available to the roots of aquatic plants.

The nutrient levels of the soils in which wild rice grows vary considerably and little information exists for specific requirements. A survey of wild rice stands in northwestern Ontario and northeastern Minnesota revealed that the soils were primarily acidic, with pH values of 5.1–7.1. The metals Fe, Mn, Zn, and Cu were present in high concentrations, at times reaching levels toxic for most terrestrial plants (Lee 1979). The concentrations of the cations calcium (Ca), magnesium (Mg), and potassium (K) were present in the order Ca greater than Mg greater than K. The nutrients with the lowest relative concentrations were nitrogen (N) and phosphorus (P). Other studies have consistently shown that P, Ca, and the metals are important factors affecting wild rice productivity (Lee 1982, Lee 1983*b*, Lee and Stewart 1984). Elements that are not nutrients, such as lead, may also be accumulated in wild rice tissues (Behan et

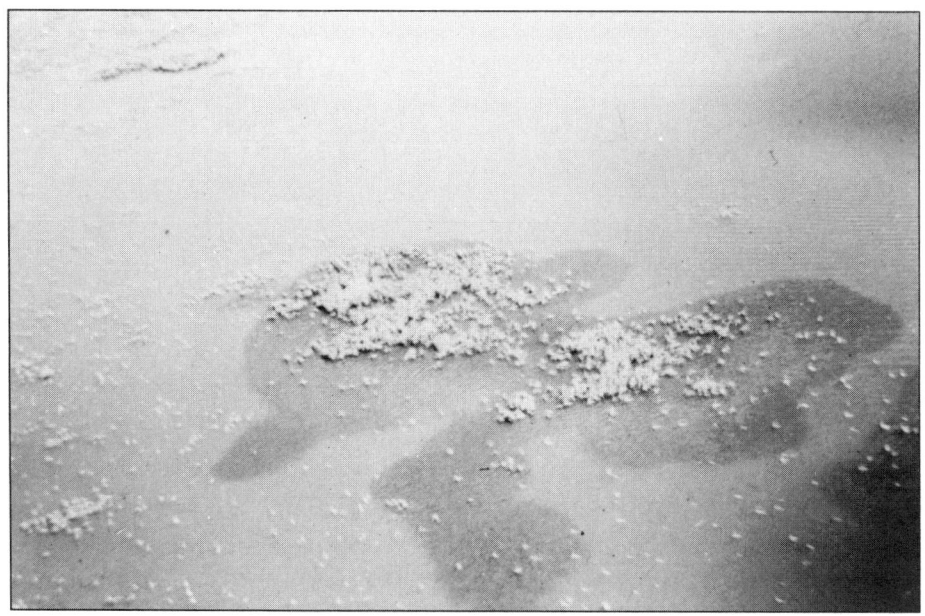

Fig. 14. Enhanced growth of wild rice within the dark, round clumps of *Potamogeton robbinsii*.

al. 1979). Lead in the soil is taken up and translocated to the shoot system, but relatively little lead enters the grains. Concentrations of lead accumulating in tissues of wild rice and other rooted aquatics do not appear to be sufficient to harm herbivorous water fowl or wildlife.

More research is required to determine the optimum levels of nutrients required in the soil for wild rice growth. Conventional soil tests, using dried soil, are inappropriate because drying the soil changes the chemical behavior of the elements. The proper method is to analyze the soils while they are still wet and to express the nutrients by volume rather than by weight. Another problem is that the levels of nutrients fluctuate by as much as a factor of 10 during the season (Lee 1983b). Based on the few studies that have been done (Lee 1982, Lee 1983b), the most productive rice areas have the following values for extractable nutrients in the fall: N greater than 2.0 g/m$^2$, P greater than 1.5 g/m$^2$, and K greater than 3.0 g/m$^2$.

## Competition with other plants

According to Moyle (1944), plants that compete with wild rice are

Fig. 15. *Eleocharis palustris*, spike rush, a perennial emergent species that commonly competes with wild rice. Most of the picture shows spike rush stems sticking up stiffly out of the water. There is a wild rice plant in the foreground with relatively wide and curving leaves.

usually perennials, including emergent, floating-leaf, and submerged species.

Emergent species that have been reported as adversely affecting wild rice are spike-rush *(Eleocharis* spp.), bulrush *(Scirpus* spp.), horsetail *(Equisetum* spp.) (Lee 1982), rigid arrowhead *(Sagittaria rigida)* (Lee and Stewart 1981), and pickerelweed *(Pontederia cordata)* (Coltas 1983). These emergent species start their growth cycle in the spring before wild rice. By the time the rice plants reach the surface of the water, the emergent plants are already well above the water surface. The underground stems rapidly spread out and send up new shoots at frequent intervals (Fig. 15). The result is that these species deplete the nutrient reservoirs of the lake bottoms. Rice plants in these infested areas are small with no tillers. In one lake near Ignace, Ont., rice plants in an *Eleocharis* stand were less than 1 g in dry weight at maturity, whereas in the same lake outside the *Eleocharis* stand, rice plants weighed up to 52 g.

Floating-leaf species, mostly water lilies *(Nymphaea* spp. and *Nuphar* spp.), watershield *(Brasenia* spp.), and bur reed *(Sparganium* spp.), can also have severe effects on wild rice. Like the emer-

Fig. 16. A dense bed of water lilies with large leaves that have spread across the water surface in the early spring and are shading out the later developing wild rice plants.

gent species, the floating-leaf species start to grow in the spring before wild rice does. If conditions are suitable, large colonies can be formed that eventually converge on one another and impede the growth of any wild rice by stopping light from penetrating the water column. Invasion of *Brasenia* into wild rice stands has recently become a serious problem in Manitoba (Stewart, unpublished observations 1985). Usually water lilies are not a problem, because they do better than wild rice in deeper water. However, in lakes such as Lake of the Woods, where the depth of the water varies by up to 0.6 m per year, the deeper water encourages the growth of these plants over wild rice. When the lake level is lowered, the lilies have become so well established that they prevent wild rice from growing in otherwise ideal areas (Fig. 16).

Submerged species that affect wild rice growth include those with finely dissected leaves, such as *Ceratophyllum* spp. and *Myriophyllum* spp. These species invade more eutrophic wild rice stands and form dense clumps that impede light from reaching the developing wild rice plants or tangle with the submerged leaves and hold the plants under water.

Blooms of free-floating algae sometimes form during the sub-

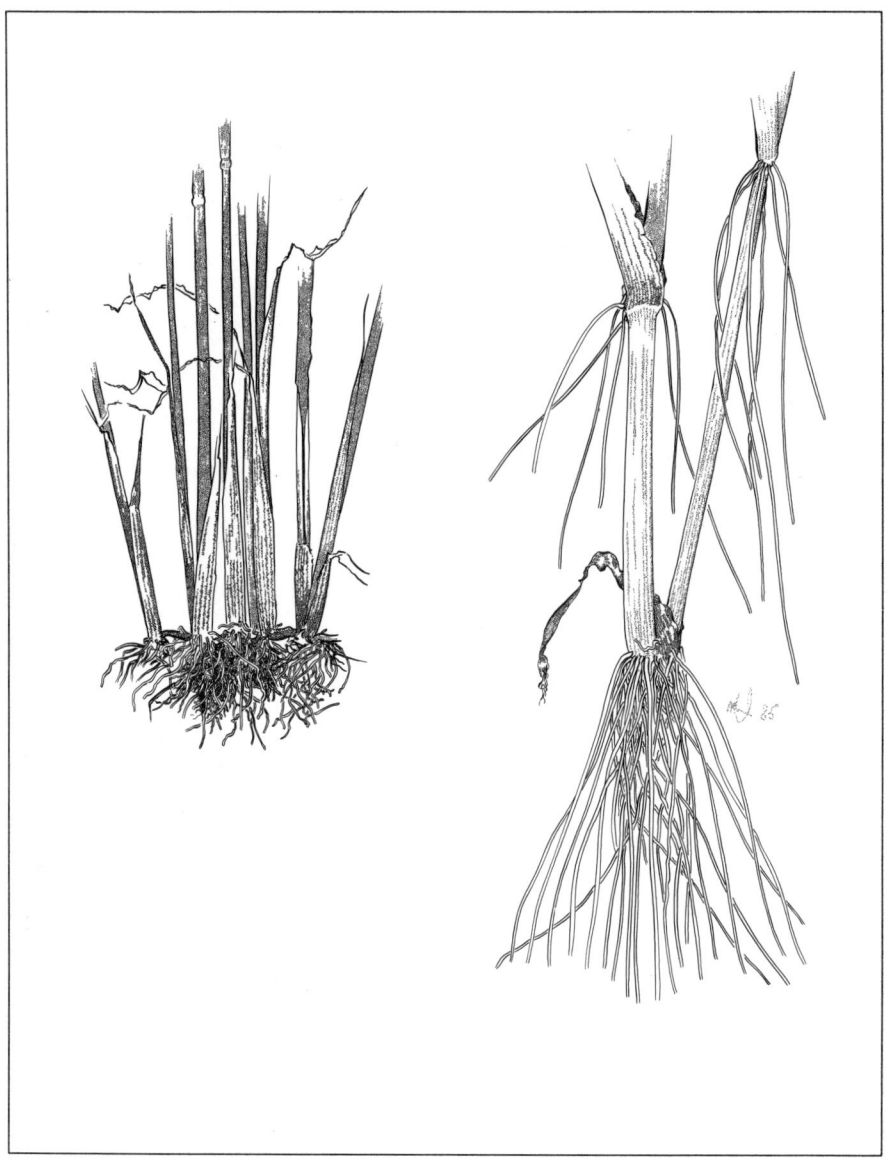

Fig. 17. Tillering, the production of many culms from one base. *(A)* Tillering in a plant that has grown in water 10–20 cm deep. *(B)* Tillering in a plant grown in water more than 50 cm deep.

mersed and floating-leaf stages of wild rice development. The algae cling to the young leaves and thus interfere with their ability to photosynthesize, eventually causing the death of the plants.

## Competition within wild rice populations

Wild rice stands are most productive with about 10 plants per square metre. In good growing conditions, tillering (Fig. 17) will occur, producing up to 50 stems per plant. Under usual, natural conditions, the densities are much greater than this, reaching over 300 plants per square metre (Lee 1982) and resulting in lower yields.

# IV
# Management of Natural Stands

Although wild rice has been seeded in lakes for centuries, little effort has been made to develop the management techniques needed for intensive cultivation. Only in recent years, attention has been focused on methods of increasing production of this crop. Today, research on wild rice is being conducted to identify suitable lakes, develop fertilizer and weed control techniques, and select strains of wild rice adaptable to various environmental conditions. Control of the water level is becoming standard practice and the efficiency of harvesting machines is being improved.

## Lake selection

A lake or river offering potential for commercial wild rice development should be 1 m or shallower for an area of at least 50 ha. In northwestern Ontario, where investigations on extending the range of commercial wild rice production are under way (Lee 1974, 1975a and b, 1979), many suitable waters exist. Trial seedings have shown that the best lakes (Figs. 18, 19) have the following characteristics:
- sufficient light penetration into the water column for tillering to occur (normally the lake or river bottom should be visible from the surface of the water)
- an organic soil with relatively high concentrations of N (1–5 $g/m^2$), P (1–3 $g/m^2$), Fe (75–200 $g/m^2$), Mg (3–15 $g/m^2$), K (3–20 $g/m^2$), and Zn 0.5–1.5 $g/m^2$).

MANAGEMENT OF NATURAL STANDS 49

Fig. 18. An aerial view of an ideal lake test-seeded with wild rice. The rice appears as a solid white streak, which indicates excellent growth.

- an absence of other competitive emergent aquatic plants.

In northern Saskatchewan, seeding is recommended in areas that have some water circulation, soft textured soils, and sufficient light penetration into the water to allow photosynthesis to occur (Orcajada 1982, Horner 1983).

Future research on potential wild rice lakes will concentrate on identifying suitable lakes with the use of satelite imagery, and on refining selection criteria based on the known requirements of wild rice.

## Seeding of potential sites

When seeding a site with wild rice, care should be taken to use a grain that is suited to the environmental conditions of the site. Research has shown (Counts 1983) that different strains of wild rice exist that are tolerant of different growing seasons, water depths, and nutrient levels. Strains of rice also vary in their yield and grain size. Eventually, improved cultivated strains will be available that have a combination of desirable traits.

Once the seed has been selected, it must be kept wet. If the seed dries out, germination will be severely impaired (Brown and

Fig. 19. A view of the water of an ideal lake test-seeded with wild rice. The rice is very dense and each plant has many tillers.

Scofield 1903). Most wild rice growers store their seed in porous bags in a lake at a depth of 2–3 m until they are ready to plant. The bags are sometimes placed in wire cages to prevent muskrats from eating the rice. Rice is also stored under refrigeration in porous bags that are periodically sprayed with water.

Seeding is generally done in the fall, the most convenient time because the seed does not need to be stored. However, some growers prefer spring seeding, claiming that the germination rate is higher after winter storage. Another approach, winter seeding (Figs. 20, 21), can be successful if the ice is not blown to the shore in the spring carrying the rice with it. Winter seeding on ice allows the use of conventional seeders. Because the operator can see exactly which parts of the lake have been seeded, a uniform seeding density can be achieved. The main problem with winter seeding is that the grain must be properly stored before it is applied.

Seeding has traditionally been done by hand broadcasting from a boat. The recommended seeding rate is about 25–35 kg wet weight per hectare. Although no seeders have been designed specifically for wild rice, a few prototypes have been constructed by modifying seeders that are used for other types of grains (Fig. 22). Aerial seeding has been tried by releasing the rice from a

Fig. 20. Winter seeding facilitates a uniform seeding rate, because the areas that have been seeded are visible.

Fig. 21. The pattern of winter seeding is still visible the following summer after the rice has grown.

Fig. 22. A seeder *(top)* mounted on the side of a boat and used to broadcast wild rice seed; close up of seeder *(bottom)*.

chute through the drop-hatch of an aircraft and by using rotary seeders that attach to the underside of the aircraft. These methods, although costly, give excellent results.

## Water depth management

For the best crops, the water of a wild rice lake must be maintained near an optimum depth of 0.6–1.0 m. Methods of water control range from keeping debris cleared from the outlets to constructing elaborate dams and trenches.

Dams constructed by beavers on the outlets of lakes are a persistent problem. These obstructions must continually be removed, either with dynamite or by hand. If culverts are used to control water levels, a screen is often placed over the opening to prevent beavers from entering the culvert. Although the beavers may still block the water flow, it is easier to remove the debris from the end of the culvert than from the middle. Another method of discouraging beavers is by using repellants. These are, however, only moderately effective and must be continually replaced.

When wild rice lakes are connected to major bodies of water, small adjustable dams may be constructed to protect wild rice stands from rising levels in the main water body.

Water drawdown can be detrimental to wild rice crops. If the water level is not raised in the fall, extremely shallow lakes may freeze to the bottom, reducing germination success the following spring. As a result, in unmanaged lakes, rice often grows only in deeper waters, producing a rice-free "ring effect" in the extremely shallow regions of the littoral zone (Fig. 23). Why wild rice does not germinate well in shallow water is not understood. Germination is not adversely affected by freezing or lack of oxygen (Svare 1960). Perhaps an arid layer develops between the surface of the sediment on which the rice seed lies and the overlying ice column, causing the rice to die because of desiccation.

## Soil management

Each year large quantities of nutrients are tied up in the wild rice straw, which decomposes slowly (Archibold and Weichel 1983, Sain 1983). Techniques such as the addition of fertilizers to wild rice lakes are now being investigated to remedy this nutrient depletion in the wild rice stands. In Nova Scotia, the addition of fertilizer to wild rice lakes was shown to increase production marginally (Melanson 1981). The most promising technique seems to be the use of

Fig. 23. Ring effect, showing wild rice growing well in the deeper regions of the lake, but absent from the shallow areas along the shore.

slow-release fertilizers that dissolve in the soil in 2–3 months. Under controlled greenhouse conditions, these fertilizers can cause a phenomenal increase in biomass production (Fig. 24). Under field conditions, the only feasible method of using these slow-release fertilizers is to use pellets that are heavy enough to sink into the bottom sediment. In this manner, the nutrients are made available to the plant roots and are not merely dissolved in the water. Recent experiments have shown (Lee 1984) that rice plants can be expected to undergo a twofold increase in dry weight in the field with these fertilizers. Future full-scale commercial applications will determine the economic feasibility of this management technique. Government policies must also be established to enable this technique to be used and to assess any environmental impact. Certainly, it will not be possible to use fertilizers in all lakes because of possible nutrient enrichment problems, but it is likely that they will be allowed for use in smaller lakes whose designated primary purpose is to grow wild rice.

## Weed control

Both mechanical and chemical control methods have been used in an effort to control competitive emergent aquatic plants.

Fig. 24. Wild rice fertilized with conventional fertilizers (a–f) and with a much more effective slow-release fertilizer (g).

Mechanical control techniques have been used in both growing and nongrowing seasons. The theory of winter removal is that the old culms that protrude through the ice supply the rhizomes in the sediment with a needed supply of oxygen. If the culms are cut off with a cutting device, the plants die from oxygen starvation. This technique works with some species such as cattail *(Typha* spp.) on sloughs in the prairies. However, when tried on spikerushes *(Eleocharis* spp.), northwestern Ontario's main problem species, it had no effect.

Summer experiments have been conducted on removing unwanted lily pads *(Nuphar* and *Nymphaea* spp.) mechanically. In these cases, cutting bars were attached to the front of airboats and

Fig. 25. Brush-on herbicide applicator, used to control competing emergent aquatic plants, mounted on a pontoon boat.

Fig. 26. Herbicide applicator pads brushing emergent weeds while wild rice seedlings are still under water.

Fig. 27. A dense mat of straw deposited on the bottom of a wild rice stand. This accumulation prevents light penetration and impedes the young seedlings from developing.

the plant's leaves were cut off. This prevents plants from obtaining energy by photosynthesis, and if pruning continues, the plants eventually die.

Herbicides have been used on an experimental basis. To date, the best success has been achieved with a "brush-on" applicator similar to that developed by Dale (1979). The applicator (Figs. 25, 26) consists of a boom containing a felt drop cloth that is saturated with herbicide by a small centrifugal pump. The applicator is attached alongside an air boat. The aim of the operation is to brush the herbicide on competing aquatic emergent plants before wild rice is in the emergent state. For this reason, application is performed in the spring. Although the technique works well, at present there are no herbicides registered in Canada for commercial use on plants that compete with wild rice.

## Thinning requirements

Crowding of wild rice plants commonly occurs in lake stands. In order to maximize production, thinning to a density of 15–30 plants per square metre is recommended during the floating-leaf stage. Equipment used for thinning consists of a series of V-shaped knives 15–20 cm apart, which are attached to a bar and pulled through the rice at a speed of about 50 km/h.

## Straw removal

Normally, the debris from wild rice plants of the previous year is swept to shore by wave action in spring runoff. However, in some cases it sinks and forms a dense mat (Fig. 27) that prevents light from reaching the young wild rice seedlings on the lake bottom, thus resulting in a severe depletion in their numbers.

Several strategies have been tried to remove this accumulated straw. Some growers drag a harrow-like apparatus through the rice bed in the spring to remove the old straw or use rotating saws that cut the rice into small pieces. Other growers have tackled the problem in the fall and use various underwater cutting bars to cut off or uproot the plants so that they will be blown to the shore. One apparatus, an aquatic tractor, is powered through the rice bed on large, ribbed wheels. The tractor not only uproots the rice plants, but also mixes the sediment, releasing nutrients from the rich, underlying layers.

# V
# Diseases and Pests

Most of the diseases and pests of wild rice *(Zizania)* have been studied for a relatively short time and are thus poorly known by comparison with those of rice *(Oryza)*. However, it is noteworthy that nearly every pathogen or pest newly reported from wild rice has previously been found on rice. This is partly a function of similarities in anatomy and physiology as well as the aquatic environment shared by these plants. It adds weight to the view that these two genera are closely related.

## Diseases

A number of disease organisms, fungi, bacteria, and viruses may be found on wild rice. Most of these pathogens attack the floating and aerial parts of the plant, causing destruction of photosynthetic or reproductive tissue and consequently reducing plant vigor and limiting seed production. Harvesters rarely have reported disease conditions, suggesting that epidemic outbreaks of disease may be relatively uncommon in natural stands or may be insignificant compared with other limiting factors. However, as increasing attention is paid to the management of natural stands and as yield expectations rise, concern for these diseases is likely to rise also.

### *Brown spot*

Fungi are responsible for several kinds of leaf damage, one of the commonest being the condition known as brown spot (Plate 1). The principal causal agents of this are the *Drechslera* imperfect states of *Cochliobolus miyabeanus* (Ito & Kurib.) Drechsler ex Dastur and *Cochliobolus sativus* (Ito & Kurib.) Drechsler ex Dastur, also referred to in the literature by their older names *Helminthosporium oryzae* and *H. sativum*. Both organisms are spread by means of airborne spores (conidia) that are produced on infected tissue as it begins to collapse. These conidia are large *(C. sativus,* up to 120 µm; *C. miyabeanus,* up to 150 µm long), dark-colored, and more or less curved.

Successful infection and subsequent disease development are favored by high temperatures (>25°C day, 20°C night), high relative humidity (>90%), and leaf wetness for periods in excess of 8 hours. Whereas *C. sativus* occurs throughout the range of wild rice, isolations of *C. miyabeanus* in Canada have been restricted so far to a few locations in southern Manitoba (McQueen 1981). *C. miyabeanus* has been the more destructive pathogen in the epidemics of helminthosporium blight, which have devastated paddies in Minnesota (Kernkamp et al. 1976). Its absence from more northerly regions, probably because of a temperature limitation, may explain the apparent rarity of destructive brown spot outbreaks at higher latitudes.

Other conidium-forming fungi that can cause brown spot symptoms are *Fusarium* spp. (Gilbert 1974) and *Dichotomophthoropsis* sp. (McQueen 1981). They have been found sporadically throughout the range of wild rice in Manitoba (McQueen 1981). Some isolates of *Fusarium* spp. were capable of causing highly significant shortfalls from the potential yield when inoculated on to wild rice by various techniques in the greenhouse and field. They appeared to be much less dependent on high relative humidity than the *Cochliobolus* spp. (Gilbert 1974).

All of the brown spot pathogens produce necrotic lesions in the host tissue, which vary in color, shape, degree of zonation, and coalescence. These lesions may be reasonably distinctive when they are the product of inoculation of plants of uniform background with a single pathogen under controlled conditions (Plate 1a–c). However, lesion type is influenced by environment and appears to depend as much on the reaction of the individual host as on the properties of the pathogen. Hence brown spot pathogens cannot reliably be distinguished in the field. Little information is available on the epidemiology and impact of brown spot pathogens in natural stands, although they are of serious concern to paddy growers. The incidence of helminthosporium blight has been reduced in paddies in

Minnesota by several applications of the fungicide mancozeb (Dithane M45); good results have also been achieved in experiments with chlorthalanil (Bravo) and propiconazol (Tilt) (Percich et al. 1985). Propiconazol is not yet registered for use in Canada.

## Leaf sheath and stem rot

Symptoms of this kind (Plate 2) are most prevalent in the shallow water of paddies but may also occur in natural stands. They may be caused by several fungi, two of which characteristically give rise to resting structures (sclerotia) in the diseased tissues. *Magnaporthe salvinii* (Catt.) Krause & Webster forms a sclerotial state, known as *Sclerotium oryzae* (Catt.), inside infected stems (Plate 2a) and in tissues of rotted sheaths. The sclerotia are black, spherical, and just visible to the naked eye. Plants with decayed stems containing the sclerotia have often been collected from natural stands in Wisconsin (Punter et al. 1984), but sclerotia and stem rot appear to be uncommon in Canada. *Sclerotium hydrophilum* Sacc. *apud* Rothert produces neither sexual nor asexual spores but relies for dispersal on the buoyant, dark brown, more or less spherical sclerotia, which may reach 1 mm in diameter. It is most often associated with a rot of the lower stem and crown (Plate 2b), especially of plants growing in shallow water. *S. hydrophilum* has been collected on a wide range of hosts in Europe, Asia, and the Pacific region as well as in North America (Punter et al. 1984). All records of this organism in Manitoba are from below latitude 51°N (McQueen 1981).

Some of the pathogens associated with brown spot may also induce sheath and stem rots. Gilbert (1974) found evidence that certain isolates of *Cochliobolus* spp. (Plate 2c) and *Fusarium* spp. (Plate 2d) were better adapted to infection and colonization of tissues of the stem and leaf sheath than the leaf blade. Conversely, the fungi that usually cause stem rot symptoms may also invade leaf blades. *Nakataea sigmoidea* Hara, the conidial state of *Magnaporthe salvinii,* and *Sclerotium hydrophilum* have been observed by Punter (unpublished) and McQueen (1981), on diseased floating leaves from many locations in southern Manitoba following incubation in damp chambers; therefore brown spot and stem rot should not be considered as distinct diseases of wild rice, but rather as forms of symptom expression induced by overlapping ranges of pathogenic organisms.

## Anthracnose

This fungal disease, caused by *Colletotrichum* sp., mainly affects the floating leaves and the lowest aerial leaves close to the water surface. The diseased tissues are typically pale yellow with orange spore masses in the center, surrounded by dark setae (Plate 3*a*). In Manitoba, McQueen (1981) found the greatest incidence and intensity of anthracnose in the more northerly stands. Once more than 5% of the leaf surface became affected, the disease could progress rapidly up the plant. Anthracnose seemed mainly a disease of senescence, which might become damaging and epidemic under environmental conditions favorable to the fungus (McQueen 1981).

## Leaf blotch

This disease (Plate 3*b, c*) usually begins along the margin or at the base of the blade. The lesions are coal black at first, but gradually become necrotic and much paler in the center as they enlarge. Tiny black fruiting bodies (pycnidia) of the causal fungus develop in these pale regions and produce vermiform conidia with several septa. The identity of the causal fungus is still in some doubt, but there are many similarities with the rice pathogen, *Phaeoseptoria oryzae* Miyake. The disease has been reported from natural and cultivated stands in Minnesota (Percich et al. 1981). In Manitoba, it appears to be very widespread in lakes on the shield but absent from those in other parts of the province. It is mainly a disease of mature and senescent aerial leaves but in some years may attack plants in the floating leaf stage. No information is yet available on the impact of this pathogen in natural stands; however, lesions may cover most of the leaf surface, which suggests that the potential for damage is great if conditions are favorable for its early establishment.

## Smut

The smut of stems and leaves (Plate 3*d –f*), caused by *Entyloma lineatum* (Cke.) Davis, is perhaps the most conspicuous disease of wild rice. Lesions on leaves and sheaths are rectangular, vein-limited, and purple black at first (Plate 3*e*). They may eventually coalesce and cover most of the surface, especially of the sheaths. Stem lesions are initially oval and shiny black. As the disease progresses, the lesions may coalesce and affect all parts of the inflorescence (Plate 3*d*). Masses of spores produced immediately below the epidermis cause the older lesions to turn lead gray (Plate 3*f*). *E. lineatum* is associated with wild rice over most of its range but has little adverse effect on the crop despite its dramatic appearance. Inoculation experiments carried out by Gilbert (1974) indicated small but insignificant reductions in seed yield.

## Ergot

Ergot has been known to occur on wild rice for centuries. Pantidou (1959) considered the causal fungus to be a distinct species, *Claviceps zizaniae* (Fyles) Pantidou. It was distinguished from *C. purpurea*, which attacks dryland cereals and grasses, on the basis of morphology, physiology, and the inoculation experiments of Fyles (1915), Wright (1942), and Brown (1948). The conspicuous part of the life cycle of the fungus is the irregularly rounded, purple brown sclerotium that develops in place of the ovary of the host floret. It usually reaches several times the width of the grain, forcing the floral scales apart and thus becoming readily detectable (Fig. 28).

Unless detached during harvest, the sclerotia may remain on the panicle after the normal grains have shattered. They float freely and will be carried to shore by wind and wave action. After exposure to cold winter conditions they can germinate in spring to give rise to capitate fruiting structures (stromata) in which new primary inoculum (ascospores) is produced. Fyles (1915) suggests that although sclerotia may germinate in water, they can only form spores when resting on mud or plant debris. The long, filiform ascospores, released when the female wild rice spikelets are open, are carried by air currents and infect the young ovaries. A sticky substance, honeydew, containing secondary inoculum (conidia), exudes from the infected spikelets. This may be transferred by splash droplets or insects to spikelets of later-formed inflorescences, which are still open. The conidia can still germinate after overwintering (Fyles 1915). The developing sclerotium acts as a sink for nutrients, thereby robbing adjacent uninfected ovaries of the resources needed for normal development. Thus estimates of shortfall from potential yield are likely to be too conservative, when based on number or weight of sclerotia present.

Ergot is widespread but occurs sporadically. It has been reported from New Brunswick, Nova Scotia, Ontario, Manitoba, and Alberta; outbreaks in 1957 at Sheffield, N. B. (Dore 1969), and in 1938 at Lac du Bois, Man. (Conners 1939), were considered economically significant. Crop failure resulting from occasional epidemic outbreaks have been suggested as a cause of famine among native communities. Ergot sclerotia accounted for 18.5% of the fresh weight of one harvest sample collected by Punter from a lake in

# Color Plates

**Plate 1. Symptoms of fungal brown spot disease.**

*(a)* Elongate, tan-colored lesions with necrotic centers and chlorotic (yellow) halos characteristic of *Cochliobolus miyabeanus* infection, 1.5×.

*(b)* Lens-shaped, chocolate-colored lesions with necrotic centers, typical of *Cochliobolus sativus* infection, actual size.

*(c)* Early stage of leaf infection by *Fusarium* species showing nearly circular, pale brown spots, actual size.

*(d)* Advanced symptoms with dark, coalesced lesions and extensive necrosis caused by *Fusarium* species, 1.8×.

**Plate 2. Leaf sheath and stem rot.**

*(a)* Numerous, tiny black sclerotia of *Magnaporthe salvinii* produced on the inner surface of a diseased stem, 2×.

*(b)* Lower node, extensively rotted by *Sclerotium hydrophilum*. White mycelium spreading within the internodal cavity and basal tissue of the leaf sheath gives rise to larger, nearly spherical sclerotia, which become dark brown at maturity, 1.5×.

*(c)* Lesion caused by *Cochliobolus* species penetrating through the leaf sheaths into stem tissue, 1.4×.

*(d)* Stem lesions caused by *Fusarium* species, 1.5×.

**Plate 3. Other widespread diseases of leaf and stem.**

*(a)* Anthracnose. Note chlorotic (yellow) diseased areas concentrated toward the margins of leaves. Orange spore masses of *Colletotrichum* species interspersed with dark setae (bristles) develop on infected tissue, actual size.

*(b, c)* Leaf blotch. Spreading lesions are often coal black at first, but become paler at the center as they enlarge. *(b)* Elongate lesions along leaf margin, actual size. *(c)* Infection at the base of the leaf blade has spread down on to the leaf sheath. Black structures breaking through the surface of the diseased area are spore-producing pycnidia of *Phaeoseptoria* species, 2×.

*(d – f)* Smut caused by *Entyloma lineatum*, actual size. *(d)* Shiny, black, raised pustules developing on inflorescence, stalks, and spikelets. *(e)* Vein-limited, rectangular spots characteristic of infection on leaf blades and sheaths. *(f)* Pustules on stems turn lead gray in the center as ripening spores separate the epidermis from the underlying tissues.

**Plate 4. Minor diseases.**

*(a)* Zonate eyespot caused by *Drechslera gigantea* with pale necrotic areas with dark concentric rings, 1.25×.

*(b)* Bacterial leaf streak. Narrow water-soaked, translucent lesions (solid arrows) later become brown and dry (open arrows). Causal bacteria are *Xanthomonas campestris* and *Pseudomonas syringae* pv. *zizaniae*, actual size.

*(c)* Bacterial brown spot resembles fungal brown spot (Plate 1), but notice creamy drops of exudate containing cells of the bacterium, *Pseudomonas syringae* pv. *syringae*, 1.25×.

*(d)* Early chlorotic streaks *(left)* and later necrotic streaks and flecks *(right)* caused by wheat streak mosaic virus, actual size.

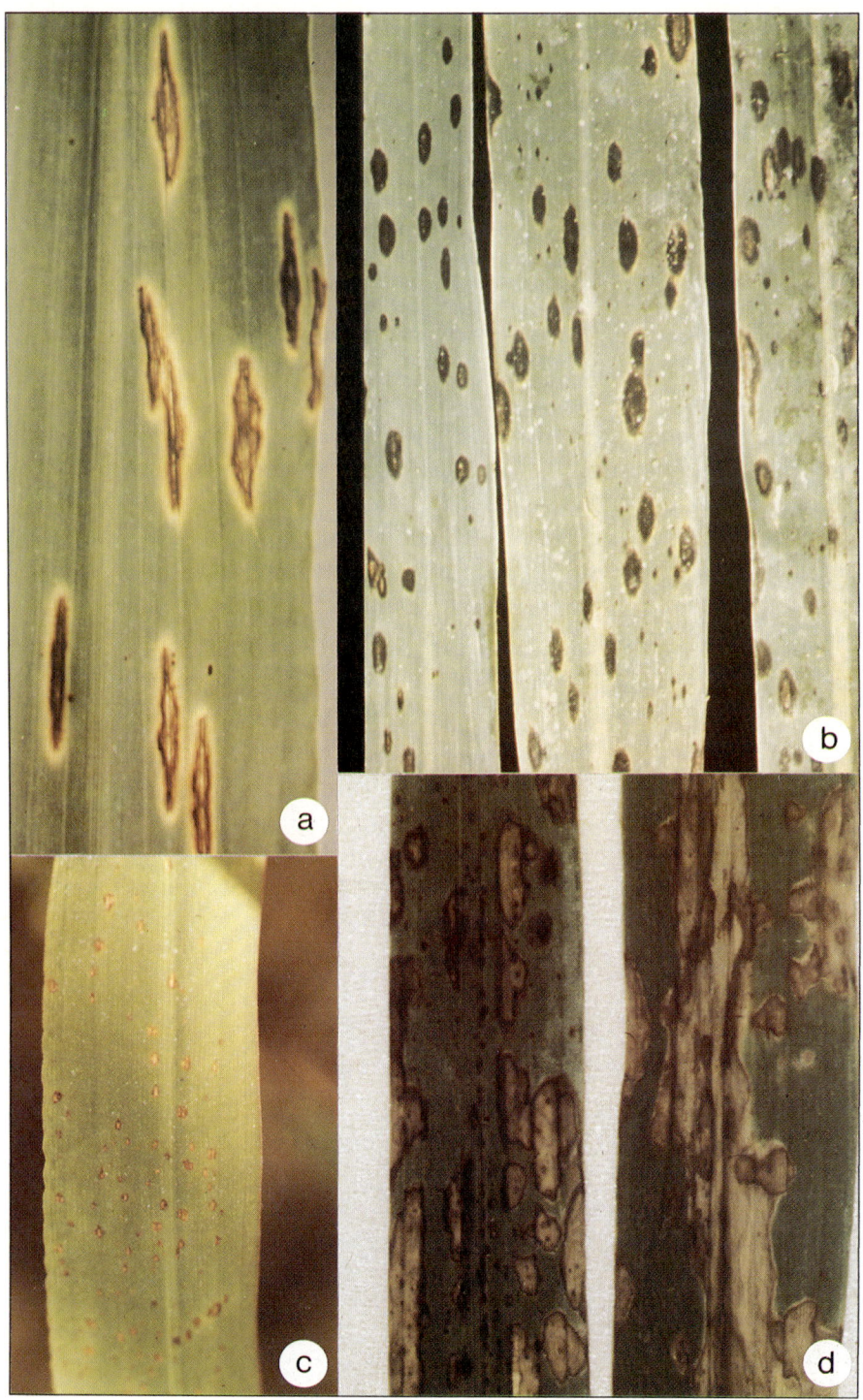

Plate 1. Symptoms of fungal brown spot disease.

DISEASES AND PESTS 67

Plate 2. Leaf sheath and stem rot.

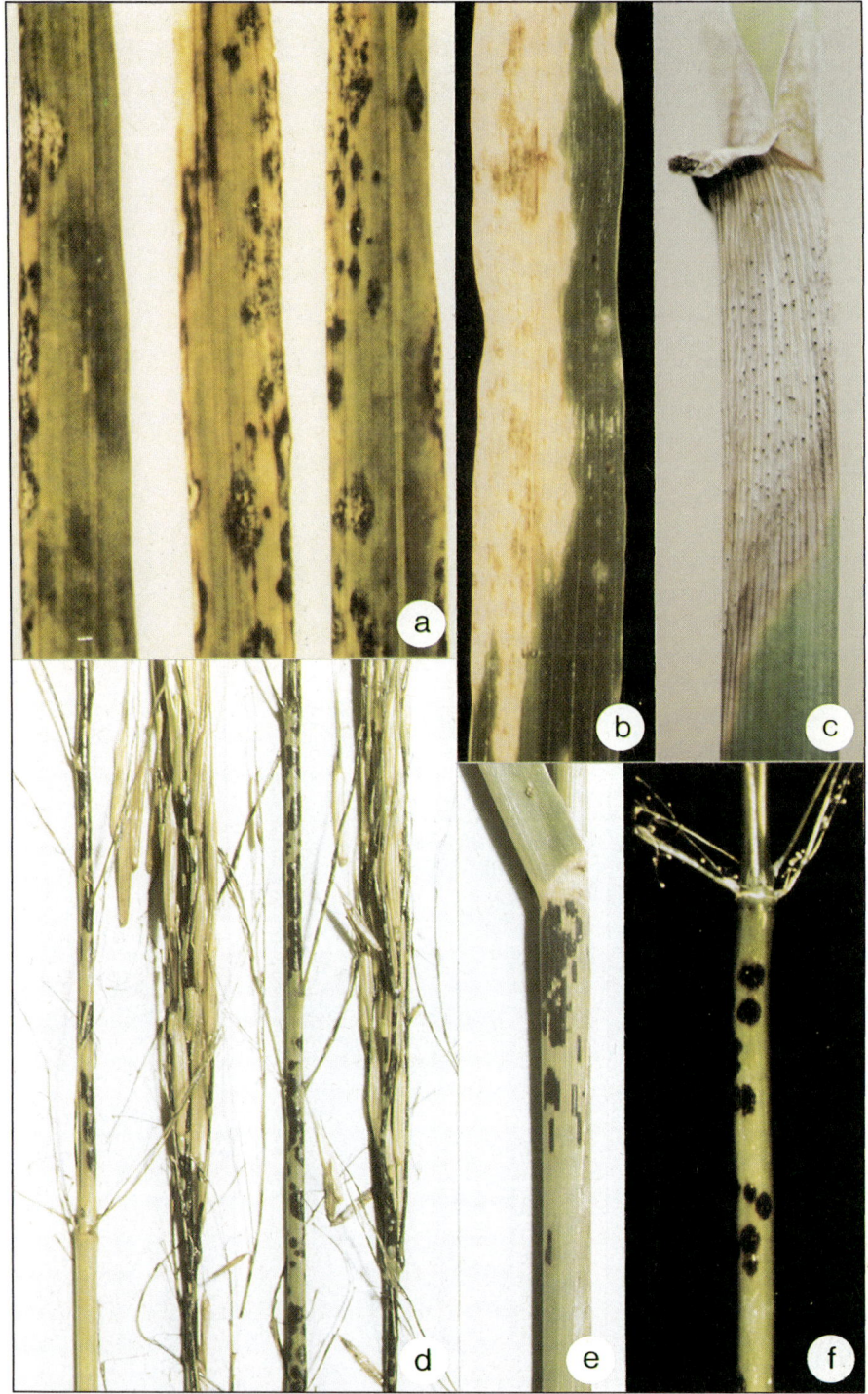

Plate 3. Other widespread diseases of leaf and stem.

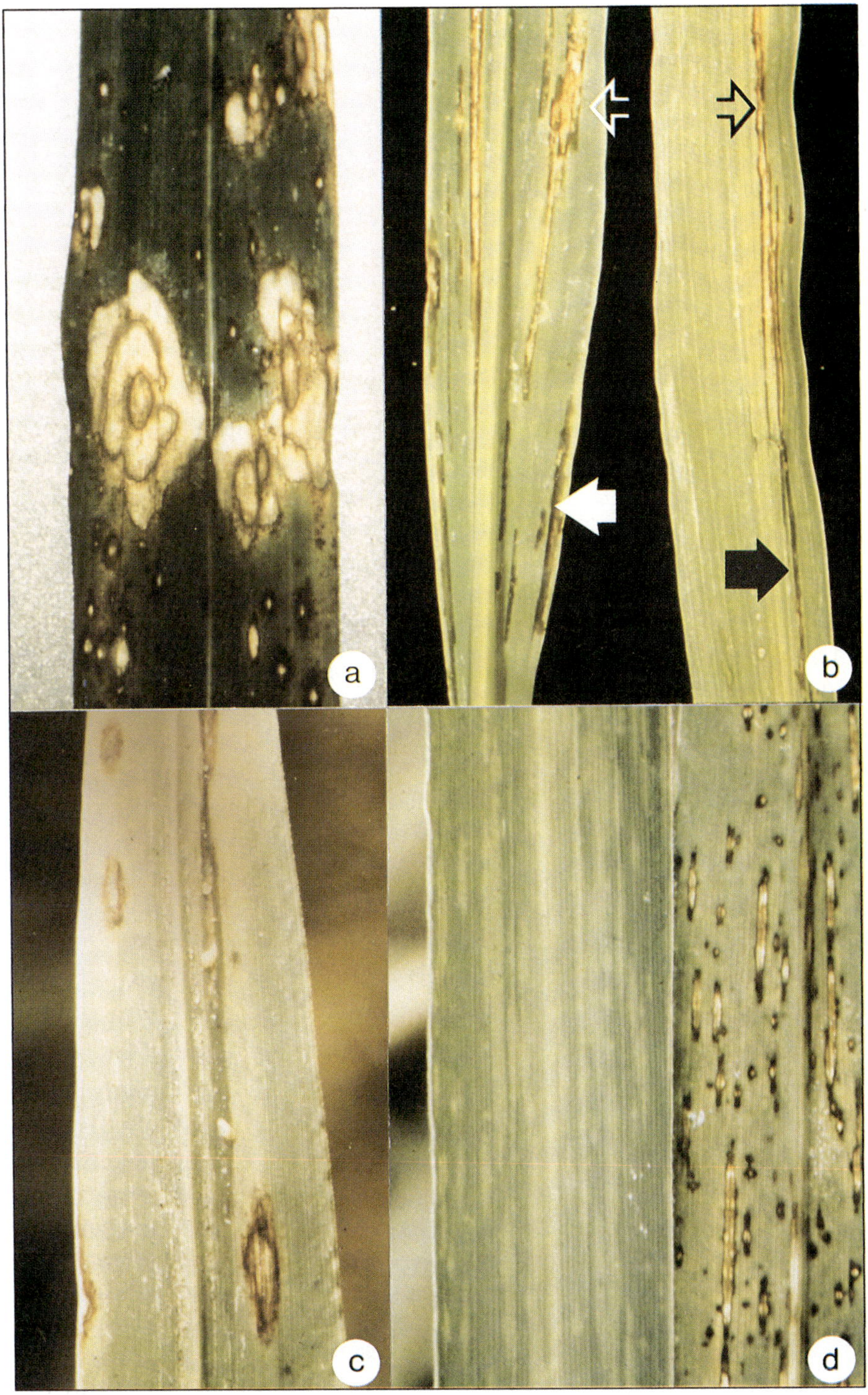

Plate 4. Minor diseases.

Fig. 28. Sclerotia of the ergot fungus, *Claviceps zizaniae*, develop in place of wild rice grains and expand the floral scales. These ergots may adhere long after uninfected grains have fallen. Approximately 0.7×.

eastern Manitoba in 1978 whereas similar samples from other lakes in the region yielded 1% or less of sclerotia. Dore (1969) mentions that ergot has been detected on northern and interior varieties and recently it was observed on southern wild rice (Darbyshire, pers. commun. 1984). It is not clear if any varietal resistance exists. Wild rice ergots, like their dryland counterparts, contain alkaloids (Taber and Vining 1960). These are poisonous to man and animals if ingested in sufficient quantity, but no published reports of poisoning by them have been seen. The separation of sclerotia from the rice maybe achieved by flotation or by screening. A limited market exists for the sclerotia themselves on account of the medicinal value of the alkaloids.

*Minor fungal pathogens*

Other minor fungal pathogens reported on wild rice in Canada are *Drechslera catenaria* (Drechsl.) Ito and *Erysiphe graminis* DC. ex Mérat (Conners 1967). In the *Index of plant diseases in the United States* (Anonymous 1960), *Diplodia oryzae* Miyake, *Doassansia zizaniae* J. J. Davis, *Mycosphaerella zizaniae* (Schw.) Lindau, *Ophiobolus oryzinus* Sacc., and *Sclerotium zizaniae* J. J. Davis are also recorded. However, the entries for *D. zizaniae* and *S. zizaniae* are both based on misidentifications (Punter et al. 1984). *Drechslera gigantea* (Heald & Wolf) Ito has been found to cause a zonate eyespot (Plate 4*a*) of wild rice in cultivated paddies and natural stands in Minnesota: this organism also occurred on various wild grasses in that state (Kardin et al. 1982). There are as yet no records of any of these fungi on wild rice in Canadian locations.

*Bacterial and viral diseases*

Two diseases attributable to bacteria and one to a virus have been described from cultivated wild rice in Minnesota. Bacterial leaf streak (Plate 4*b*) may be caused by *Xanthomonas campestris* pv. *translucens* Dye, *X. campestris* pv. *cerealis* Fang et al., and *Pseudomonas syringae* pv. *zizaniae* Bowden and Percich. The narrow, water-soaked, translucent lesions eventually become brown or black and dry (Bowden and Percich 1983*b*). Bacterial brown spot lesions (Plate 4*c*) are difficult to distinguish from those of fungal brown spot; they tend to be more irregular or diffuse, have a translucent central slit or spot, and lack any chlorotic halo. The causal agent is *Pseudomonas syringae* pv. *syringae* van Hall (Bowden and Percich 1983*a*). Wheat streak mosaic virus (Berger et al. 1981) is the only viral pathogen currently known to attack wild rice. The symptoms are shown in Plate 4*d*. These bacterial and viral pathogens are rare or unknown in natural stands and none of them has been found in Canada.

## Pests

### Invertebrates

The most comprehensive study of insects in natural stands of wild rice was carried out by Melvin (1966). By far the most damaging species is the riceworm, *Apamea apamiformis* (Guenée). Early reports of caterpillars attacking developing grains (Jenks 1901, Fyles 1920) are almost certainly attributable to this insect, a noctuid moth whose identity and life cycle were clarified by MacKay and Rockburne (1958). The adult moths emerge and mate early in July as wild rice panicles begin to appear. One of their principal food sources is the nectar of the common milkweed, *Asclepias syriaca* L. (Peterson et al. 1981). Female moths lay their eggs in the female spikelets during the short period when the lemma and palea part to expose the receptive stigmas. Twenty-five eggs may commonly be laid in a single spikelet, and much greater numbers are reported. They are about 0.5 mm in diameter, creamy white at first, darkening to gray within a week or so. At this latter stage, just before hatching, the egg masses are readily visible in silhouette if the spikelets are held up to the light or cleared (Fig. 29).

The larvae, which are greenish when first hatched, become browner with age. The darker longitudinal stripes, characteristic of all but the first and last instars, provide a degree of camouflage against the background of the maturing plant. Larvae in the early instars feed mainly on the developing ovaries and floral parts. After consuming the contents of a spikelet, they cut an exit hole through the outer scales and migrate to other parts of the panicle (Fig. 30). The silk threads that they spin tend to bind the panicle together and also allow the larvae to be blown from one plant to another. Older larvae may continue to feed on the kernels (Fig. 31) but may also move into a leaf sheath, or sometimes into the upper part of the stem. Late instar larvae pass the winter in soil. Those on plants near the water's edge may be able to crawl ashore, whereas others are rafted there in broken or uprooted stems or on mats of debris. The overwintered larvae undergo a final molt and pupate in June.

Riceworm seems to occur wherever wild rice grows naturally or in cultivation. Hammond (1958, 1959) found large populations in eastern Ontario during 1957 and 1958, with destruction of entire stands at some locations. Steeves (1952) reported similar damage in Manitoba. Riceworm populations were also high in Manitoba in

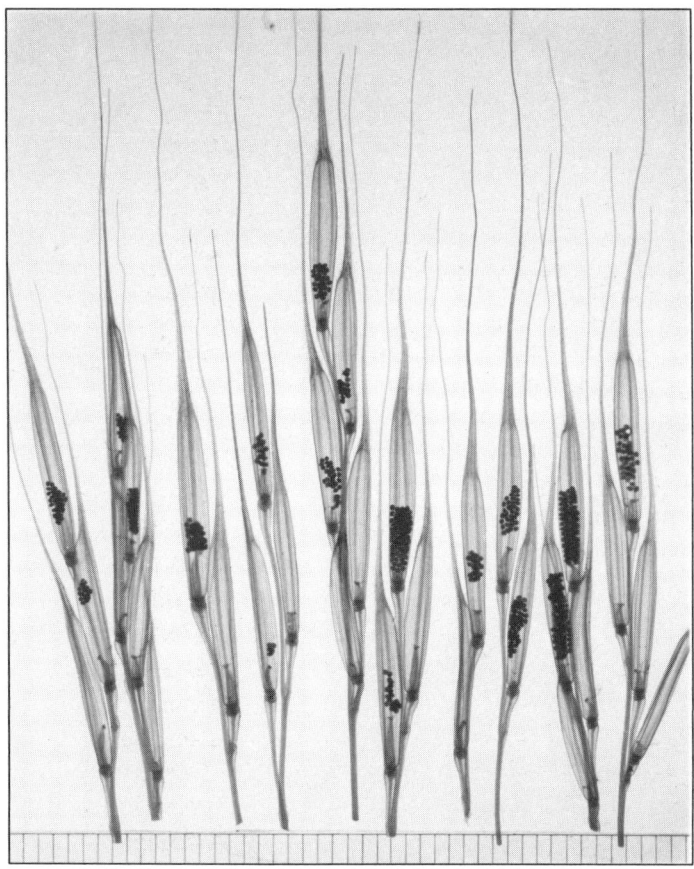

Fig. 29. Egg masses of the riceworm *Apamea apamiformis* inside female spikelets. For this picture the opaque hulls were "cleared" in the laboratory. In the field they may be seen in silhouette by holding the spikelets up to the light. Approximately 3×.

1957, declined to negligible levels by 1960, and were rising again by 1963 (Melvin 1966). The extent to which these cyclical changes were the cause or effect of fluctuation in wild rice crops is not clear. Parasitism does not seem to be a controlling factor, because only one ichneumonid parasite, *Gambus bituminosus* Cush., was recorded from late instar larvae (Melvin 1966). Experiments by Peterson et al. (1981) indicate that populations of one larva per panicle can be expected to cause a yield loss of not less than 10%, since each larva is able to consume seven or eight kernels. Riceworms are controlled effectively by application of malathion (1 kg/ha) within a few days after hatching. Some other insecticides also give good control, but spore suspensions of *Bacillus thuringiensis* (Dipel)

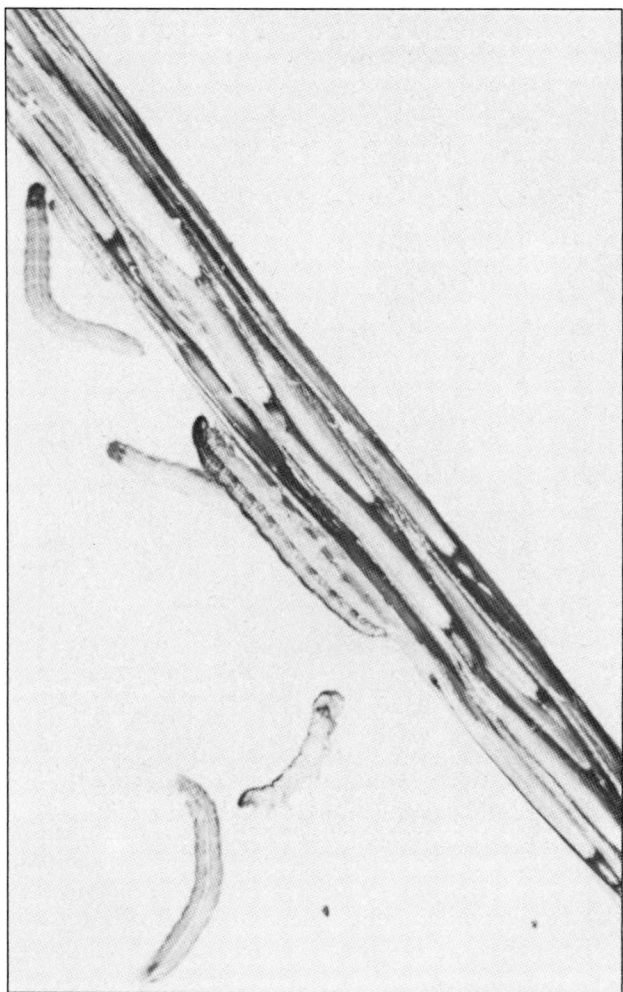

Fig. 30. In some years riceworms (larvae of night-flying moth, *Apamea apamiformis*), severely damage the wild rice heads. The late instar larvae shown here are migrating over the outside of an infested wild rice inflorescence. Approximately 2×.

do not. By doing an egg count and estimating spraying costs, potential yield, and potential value of the product, the producer can determine the economic feasibility of spraying for riceworm control.

The rice stalk borer, larva of the pyralid moth *Chilo plejadellus* Zincken, may often be found in the lower portions of the stems near the root crown. This organism also overwinters in the larval state within the stems of wild rice and many other aquatic plants. Pupation occurs in these stems and adults emerge in mid-June.

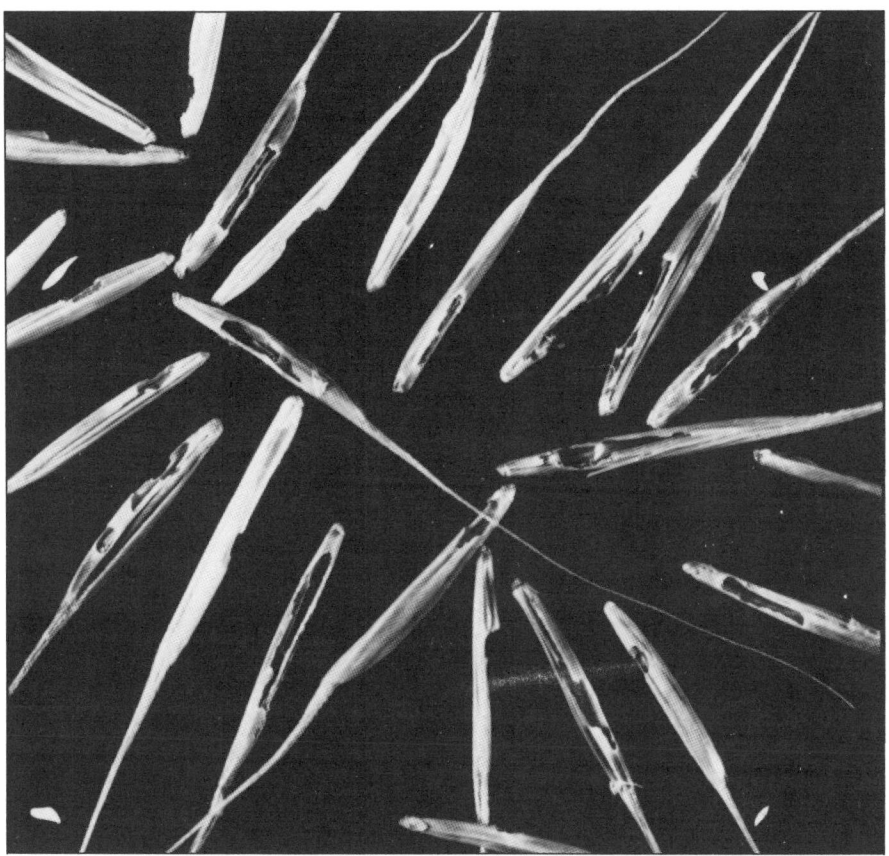

Fig. 31. Damage to wild rice grains caused by larvae of *Apamea apamiformis*. Approximately 2.5×.

The oval, creamy white eggs are laid on floating leaves in overlapping rows. They incubate for about a week, turning orange just before hatching. After feeding briefly on foliage, the whitish first instar larvae bore into the lower part of the stem, where they feed on the pith diaphragms and the tissues lining the stem cavity. Older larvae, though superficially similar to riceworms, may be distinguished by the cervical shield, a darker band immediately behind but separate from the head capsule.

The rice stalk borer is widely distributed along with the riceworm in areas where wild rice is prevalent, for example Manitoba (Melvin 1966), Ontario (MacKay and Rockburne 1958), and Minnesota (Peterson et al. 1981). Peterson et al. (1981) found only 5–10% yield reduction and minor loss of quality even when all stems were infested; few lodged plants or "white heads" were observed in

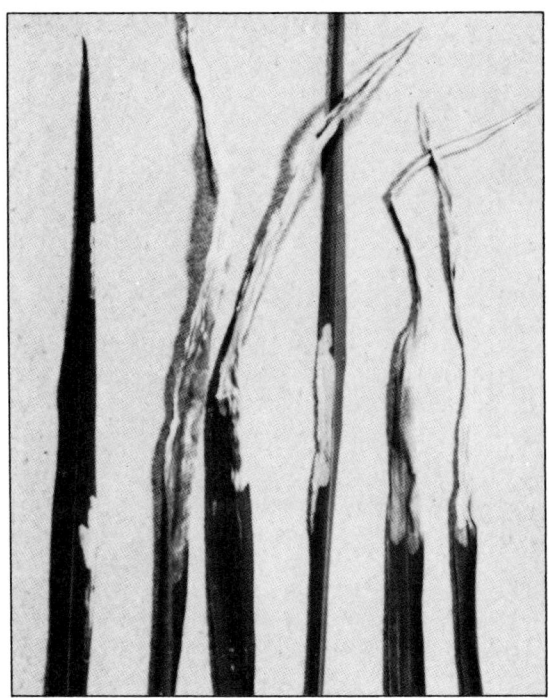

Fig. 32. Damage to leaves by the leafminer insect. *Phytobia incisa,* actual size.

this connection. None of the five types of injury, which are common on white rice, is considered of much importance on wild rice. Both Melvin (1966) and Peterson et al. (1981) reported the occurrence of the braconid parasite, *Chelonus knabi* Vier., in overwintering larvae. The incidence of parasitism may be sufficient in some circumstances (Sargent 1976) to exercise a significant degree of control.

Although leafminers are common in natural stands, they have rarely been identified. *Phytobia incisa* (Meigen) attacked all varieties of wild rice in experimental plots in Ottawa in 1962 (Dore 1969), whereas *Hydrellia griseola* (Fallèn) and *H. ischiaca* Loew are widespread and abundant in Minnesota (Peterson et al. 1981). These dipteran larvae feed on the photosynthetic tissue of the floating leaves, producing areas in which only the clear epidermal layers remain and causing chlorosis of the distal portions of the leaves (Fig. 32). Individual plants may be severely weakened or killed, but damage to large stands is rarely a matter of economic concern. Larvae of *Eribolus longulus* (Loew), the wild rice stem maggot, may also feed inside leaves. They were occasionally observed in the

Whiteshell area of Manitoba (Melvin 1966), but the damage was insignificant. This insect may sometimes prevent the emergence of the inflorescence if it enters the stem during the "boot" stage (Peterson et al. 1981).

Local damage may be caused by aphids when they become abundant. Those associated with wild rice mainly belong to the genus *Rhopalosiphum*. *R. niger* Richards is recorded from Ontario (Richards 1960) and Manitoba (Robinson and Bradley 1968); this species overwinters on hawthorns *(Crataegus* spp.). Melvin (1966) noted predation by a hover fly and a lady beetle as well as parasitism by the braconid, *Aphidius obscuripes* Ashmead. *R. nymphaeae* (L.) has been collected in Ontario on wild rice and other aquatics and in Manitoba on *Lemna* spp. (Richards 1960). Adams (1945) reported very large populations of *R. prunifoliae* Fitch causing much foliar and head injury of wild rice at Jemseg, N. B. This record should probably be referred to *R. padi* (L.), a cosmopolitan species known to occur on wild rice in Minnesota (Peterson et al. 1981). *R. nymphaeae* and *R. padi* both overwinter on *Prunus* spp. The distribution of all these aphids must be somewhat dependent on the availability of suitable overwintering hosts.

Other insects encountered by Melvin (1966) on living plants in Manitoba include a pyralid moth (possibly a *Catoclysta* sp.) and two leaf beetles *(Donacia aequalis* Kirby and *D. magnifica* LeConte). Damage caused by these insects was of little or no consequence. Some additional arthropods that occur on wild rice in Minnesota, but which have not been reported in Canada, are midges *(Cricotopus* sp.), two rice water weevils *(Lissorhoptrus oryzophilus* Kuschel and *L. buchanani* Kuschel), an aphid *(Macrosiphum avenae* Fabr.), the aster leafhopper *(Macrosteles fascifrons* (Stol)) (Peterson et al. 1981), and an eriophyid mite *(Aceria tulipae* Keif.) (Berger et al. 1981). The last two are known vectors of certain viruses. Terrell and Batra (1984) have observed bees *(Bombus vagans* Smith, *Dialictus imitatus* Smith) and syrphid flies *(Toxomerus politus* Say) gathering and feeding on pollen of *Zizania aquatica* in Maryland. This confirms an earlier report by Thieret (1971) of bumble bees gathering pollen on *Z. palustris* in northern Minnesota. In no case were the insects visiting female spikelets; their activities must therefore be considered predatory and most unlikely to result in successful pollination.

Insects are occasionally found in association with wild rice during processing and storage. Melvin (1966) detected small numbers of the dermestid beetle, *Perimegatoma vespulae* Milliron, and a tineid micromoth among chaff and broken kernels in a processing plant in Manitoba. Insects in a lightly infested sample of grain stored in Manitoba were identified by Dr. N. J. Holliday as the granary weevil, *Sitophilus granarius* L., and a spider beetle, *Ptinus villiger* (Reitter).

The nematode *Radopholus gracilis* (De Man) Hirschmann has been recovered from the cortex of roots of wild rice in Ontario (Sanwal 1957). Although well adapted to the aquatic environment and parasitic on the root tissues, this worm apparently causes very little damage to the plants. Another nematode species, *Hirschmaniella pisquidensis,* was recently described from damaged roots of wild rice growing in Prince Edward Island (Ebsary and Pharoah 1982).

Crayfish have caused some local problems in wild rice paddies in Minnesota (Noetzel, pers. commun.), but have never been worthy of note in Canadian stands.

## Vertebrates

The carp has been blamed for the decline of wild rice in Rice Lake and other locations in Ontario (Nickels 1952). This large, coarse fish, introduced into many parts of North America, tends to frequent warm, shallow waters. Because of its violent thrashing movements, the carp can dislodge loosely rooted aquatic plants such as wild rice, especially during the floating-leaf stage. Large populations of carp may uproot large areas of susceptible plants, including wild rice, wild celery, and water milfoil (McCrimmon 1968), and cause increased turbidity, which in turn restricts photosynthesis. Powles et al. (1983) found that the digestive tracts of carp, regardless of age or season, usually contained some plant material but this represented less than 10% of the diet by volume. Seeds of wild rice were frequently recovered even in winter. Although the proportion of plant material in the digestive tract was usually less than the proportion of debris, Powles et al. (1983) considered this small degree of herbivory was deliberate rather than accidental. However, they concluded that populations of carp in the Bay of Quinte watershed did not currently constitute a threat to aquatic vegetation. Lee and Stewart have observed snapping turtles grazing on the submerged leaves of wild rice in Manitoba and Minnesota.

Birds of many kinds, and especially waterfowl, are attracted to wild rice. It is recognized as one of the principal autumn foods of ducks in the eastern United States (Chambliss 1922) and has been planted in many areas as a lure for them. McAtee (1917) lists 13 species of ducks and 2 of geese known to feed on wild rice. The grains are the most commonly consumed part of the plant. Martin and Uhler (1939), in a study of stomach contents of 18 duck species from across the United States and Canada, found that wild rice constituted about 2% by volume of all the material examined and ranked ninth in the list of food plants; in eastern Canada (including Manitoba) and the eastern region of the United States it ranked third, making up almost 10% and 5%, respectively. In addition to the grains, ducks may also eat the young shoots and wood ducks eat the spikelets; wild geese consume stems and leaves of mature plants (McAtee 1917). Although wild rice stands may represent the preferred breeding habitat for certain ducks (Peden 1977), resident populations are not usually high enough to result in significant damage to the crop. Local damage may sometimes be caused when migrants arrive before the harvest is complete. However, the main flocks usually appear after harvest, using the stands as staging areas and eating only fallen kernels.

Among other birds that eat wild rice grain are bobolinks, sora rails (Chambliss 1922), sparrows (Dore 1969), and red-winged blackbirds. Blackbirds, in particular, may descend on a ripening stand in large flocks and cause serious losses, both by eating the kernels in the "milk" stage and by knocking off ripe ones that have not yet shattered. Attempts to control blackbird damage in paddies by use of the chemical repellent methiocarb have been generally unsuccessful (Moulton and Weller 1978).

A variety of mammals graze on the young aerial shoots of wild rice when the opportunity arises. Moose, deer, and domestic cattle are implicated by Steeves (1952). Damage from such activity is usually local and minor. Muskrats are of more concern in many areas. Dore (1969) wrote that they were blamed for failures in establishing wild rice in Prince Edward Island and Martin and Uhler (1939) reported destruction of up to 50% of a stand in the upper Mississippi River area before maturity. Small stands and new plantings in the Ottawa Valley have been eliminated because all young aerial shoots were clipped off at water level; larger stands are somewhat less vulnerable. Beaver can also eliminate wild rice by raising water levels through dam construction.

# VI

# Harvesting and Processing the Grain

The precise date when wild rice was first used as a food is not known. It would coincide closely with the time when aboriginal man moved eastward into central North America some 10 000 years ago, after the Pleistocene glaciation. Nomadic man, omnivorous and subsisting on whatever he could gather from the land, would quickly learn the food value of wild rice grains and settle down where a supply was available.

The eminent ethnologist, A. E. Jenks (1901), considered that, "no other section of the North American continent was so characteristically an Indian paradise as far as spontaneous vegetal food is concerned, as was this (wild rice) territory in Wisconsin and Minnesota."

## Harvesting in natural stands

By the time the first historical accounts were written, starting with Marquette's journal about 300 years ago, some tribes had long lost their nomadic habits and developed a fully sedentary way of life that depended on the wild rice harvest (Lips 1956).

The first European explorers in the area, the early traders, missionaries, and pioneering settlers, quickly learned the food value of wild rice from the Indians. In 1766, Carver (*in* Jenks 1901), recognizing its provident value, confidently predicted that "in future periods it will be of a great service to the infant colonies as it will afford them a present support until in the course of cultivation other supplies may be produced."

Alexander Henry, in his *Travels* written in 1775, believed that without the large quantity of wild rice he obtained in the Lake of the Woods area, his voyage northwestward beyond the Saskatchewan River could not have been completed.

The "cultivation" of wild rice was recommended as early as 1828 by T. Flint, who stated in his *Geography and history* "it is astonishing, amidst all our eager and multiplied agricultural researches, that so little attention has been bestowed upon this interesting and valuable grain. It has scarcely been known, except by the Canadian hunters and savages, that such a grain, the resource of a vast extent of country, existed. It surely ought to be ascertained if the drowned lands of the Atlantic country will grow it. It is a mistake that it is found only in the northern regions of the Mississippi Valley."

Flint's foresight in 1828 was accurate in predicting that wild rice would grow in many more places than its current distribution indicated. Today, wild rice has attained the status of a cultivated cereal, to be sown, cropped, and exploited on an agricultural scale.

## Early methods of harvesting and processing

Menard, a seventeenth-century Jesuit missionary, is believed to have recorded the first written description of a wild rice harvest.

"There is in that country (North America) a certain plant, four feet or thereabouts in height, which grows in marshy places. A little before it ears, the Savages go in their canoes and bind the stalks of these plants in clusters, which they separate from one another by as much space as is needed for the passage of a canoe when they return to gather the grain. Harvest time having come, they guide their canoes through the little alleys which they have opened across the grainfield, and bending down the clustered masses over their boats, strip them of their grain. As often as a canoe is filled, they go and empty it on the shore, into a ditch dug at the water's edge. Then they tread the grain and stir it about long enough to free it entirely of hulls, after which they dry it, and finally put it into bark chests for keeping." (From *Jesuit relations, 48:121–123,* as quoted in *Harrowsmith* 19:51, May 1980.)

Not all tribes bound the rice into bundles, but in general, the harvesting of wild rice by Indians was entrusted to the older members of the tribe. One person sat in the front of the canoe to propel

Fig. 33. Traditional two-stick method of harvesting wild rice.

it with a paddle or forked stick; another person sat in the back to draw the stalks in over the side with one stick and tap the ripe grains off into the canoe with another (Fig. 33). Unripe grains that adhered to the panicles were left for a later harvesting. Two or three gatherings were made during the harvest season, extending over a period of 2–5 weeks.

The Indian method of dehulling and separating the kernels from the chaff was simple but practical. The freshly harvested grain was first dried by spreading it out on either skins, bark, or an expanse of flat rock in the sun, then stirring it from time to time. The dried grain was then stored until it was convenient to parch it.

For parching, a moderate amount of the grain, not more than 10 kg, was put into a large bowl hung over an open fire. The grain had to be stirred continuously to prevent it from burning. Just enough heat was applied to swell the starch grains and force apart the tightly closed hulls. After the seed cooled, it was placed in a shallow hole on hard ground and "danced" under foot. The material so threshed was tossed into the wind, and by this simple method of winnowing, the grain was separated from the chaff and debris (Fig. 34).

Fig. 34. Traditional methods of preparing wild rice for food.

The methods of processing green wild rice and the kind of primitive equipment used differed among tribes. Manufactured utensils were introduced during the past 70 years, and the newer technologies have been incorporated into the original methods of finishing the rice. Excellent reviews of the activities and methods associated with the historical harvesting and processing of wild rice are given by Steeves (1952) and Lips (1956).

## Current methods of harvesting and processing

Mechanical harvesting versus traditional handpicking is a controversial issue in certain wild rice harvesting areas of Canada; for example, in Whiteshell Provincial Park in Manitoba, handpicking is preferred by some native groups. Mechanical harvesters are permitted on crown lands in all Canadian provinces, but they are still restricted in Minnesota and Wisconsin. On tribal lands, the Band Council decides whether handpicking or machine harvesting will be permitted, thereby ensuring access to ricing by all members of the band. Tradition, family interactions, and aesthetics are the main tribal reasons why some band members still prefer handpicking. Mechanical harvesting involves fewer people and larger areas can be harvested in less time.

Today much of the wild rice is harvested from natural stands by machines rather than the traditional canoe and flail methods. The principle involved in machine harvesting is the same as that of the aboriginal method; that is, it is a multiple-pass procedure. Both techniques aim to shake off the ripe grains and catch them before they fall into the water, and at the same time, to leave the stems and panicles undamaged, allowing immature seeds to ripen. With the increasing use of larger and more efficient machines, wild rice stands of varying densities are being harvested profitably over more extensive areas than previously. Most operators of mechanical harvesters believe the use of machines more than doubles the yields of wild rice grain obtained by hand harvesting. Even so, it is estimated that only a small portion (10–30%) of the total grain from natural stands is ever harvested, because of the variations in the density and distribution of natural stands and because of modest harvesting efficiencies coupled with the vagaries of nature.

Mechanical harvesters have been in use for many years and have gone through much development. Early models used small revolving reels to beat the grain from the plants (Fig. 35). These models were powered by outboard motors and tended to bog down

Fig. 35. An early mechanical harvester that used small, revolving reels to beat the grains from the plants.

in very thick rice stands. Today, most mechanical harvesters use a speed head, a tray-like device 1–6 m long, attached to the front of a boat (Fig. 36) powered by an air-prop engine (Fig. 37). When the rice stalks hit the edge of the speed head, the ripe grains fall in. For mechanical harvesters, the speed is critical; the recommended speed is 12–15 km/h. At slower speeds the rice tends to be knocked into the water rather than into the speed head, whereas at faster speeds the seed panicles are broken off. Mechanical harvesters operated at the appropriate speed allow a wild rice stand to be harvested six or seven times during the harvest season and can result in yields of up to 350 unprocessed kilograms per hectare.

The actual time of harvest varies with local conditions, but harvesting usually starts in mid August and continues into October, depending on the number and ages of tillers and the climatic conditions. The highest yield of processed grain is obtained when harvesting crops in which about one third of the grain on the panicle is dark and the kernels have a firm dough consistency, containing about 35–40% moisture. These conditions usually occur after a number of mature seeds have shattered from the main stem, a situation important in the reseeding of natural stands. The decision to

Fig. 36. A tray-like speed head on a mechanical harvester.

Fig. 37. A mechanical harvester with speed head in front and air-prop engine being used to propel the boat.

harvest at a particular time is influenced by such climatic events as hail, high winds, and frost.

The potential yield from a stand may be estimated by weighing the grain harvested from small unit areas (quadrats) and multiplying up to commercial areal units (hectares). Measuring the potential available rice involves counting a number of plant variables such as mature seeds, empty hulls, and seed scars per head. Because wild rice plants are rarely distributed uniformly, these estimated yield values are only approximate. Such exercises can give values of 55–682 kg/ha.

Harvested green wild rice has a high moisture content (about 40%) and must be prepared for parching before biological heating and fermentation affect the quality of the unprocessed rice. Moist starch and nutrients from broken and squashed grains provide a rich medium for microorganisms, and if such conditions are not controlled, the final product can spoil. The perishable product must be shipped rapidly by truck, rail, or aircraft to the processing plant (Fig. 38). Finely woven burlap, fiber-glass bags, or cotton sacks are used to ship wild rice, but when a bumper crop is in danger of spoiling, any kind of open container can be used.

Grain required for reseeding should be stored wet until needed. The viability of wild rice is reduced if grain dries to less than 30% moisture (Elliott 1980). Grain for spring sowing is routinely submerged in shallow lakes over the winter.

When the green rice is delivered to the processing plant, it is spread out to cure, often in windrows, over a clean flat surface up to a depth of 50 cm (Fig. 39). Curing, or fermenting, involves heat and moisture transfer, large numbers of microorganisms, and both plant and microbial respiration. This complex chemical and biological process allows immature grains to ripen. Many grains change from soft to firm and become a characteristic dark brown or black during this period. Many processors believe fermentation is necessary for flavor development.

Curing generally lasts from 4 to 7 days. Because of imbalances between harvest capacity and processing capacity, fermentation may be extended up to 3 weeks, although losses in green rice weight will result from increased respiration (Oelke et al. 1982). The batches of grain are sprayed periodically with water and turned over by machines or stirred with a pitchfork at least once a day to expose all grains to air and prevent souring, decay, and dry weight loss.

Early parching ovens were made from empty oil drums and were turned by a hand crank. Today, many parching ovens are large, 1–2m

Fig. 38. Sacks of harvested green rice being loaded on a float plane for rapid transportation to a processing plant.

Fig. 39. Green rice spread in batches on a flat surface to start the curing process. Seed is sprayed with water and turned over daily to prevent souring and decay.

Fig. 40. Parching ovens. Wild rice grains *(right)* being shoveled into an open oven; closed oven *(left)* in operation with steam rising. The inner drum rotates while heat is applied by gas to the bottom of the parcher.

cylindrical structures that rotate over gas-fired jets. They usually handle up to 270 kg of wild rice per batch (Fig. 40). Parching usually lasts 1.5–3 hours, depending on the initial moisture content of the seed. Care must be taken not to overheat the grains or they will pop. Even heating is achieved by continuously rotating the ovens, avoiding overloading, and allowing the temperature to rise gradually until it is about 350°C. As long as steam issues from the ovens, the rice is still drying. After parching, the grain is dumped from the ovens onto a clean surface where it is allowed to finish drying and to cool. Parching makes the hulls brittle so that they break and even pulverise, which helps with their removal.

The cooked kernels should be firm and enclosed more or less loosely within their hulls; extraneous leaf and stem fragments or chaff become dry and all caterpillars or other previously living organisms are dead and shriveled. The parching process reduces the moisture content to less than 10% and gelatinizes the starch. The parched rice can either be held in storage in this dry state or be sent into the next stage, the hull-cleaning process. In some modern processing plants, the whole parching process is automatic, with thermally controlled, continuous flow equipment.

Fig. 41. Dehulling machines remove wild rice chaff from the grains. Rubber-covered flails inside rotate in opposite directions and scrape off the chaff, which is blown to an exterior disposal pile.

Dehulling (or threshing) the grain is done in large cylindrical containers about the same size and shape as the parching ovens (Fig. 41). The dehuller containers are fixed in a stationary position and internal flails or cylinders are covered with rubber and rotated at predetermined speeds. The kernels shell out readily, and threshing must not be done severely or the grain may break. This dehulling process is identical in principle with the beating or dancing of the rice that was originally performed by the Indians, and it is the same as the one used in modern farm threshing machines. Although there is always some breakage of long kernels, with care the proportion can be kept small. A jet of air forced into the dehuller during the threshing carries out the empty hulls and any light extraneous matter, leaving behind the clean smooth kernels and the heavier foreign matter.

After the dehulling process, some processors remove a portion of the outer layer (or pericarp) of the wild rice kernels, a process called scarification. The scarifier is a closed cylindrical container in which several rubber paddles rotate at speeds that influence the degree of scarification. The procedure shortens the cooking time, which is important when wild rice is mixed with white rice.

HARVESTING AND PROCESSING THE GRAIN 91

Fig. 42. (*left*) The screening machine removes any foreign material (stones and such things) that might remain after the dehulling process.

Fig. 43. (*below*) The dimpler machine separates the broken kernels from the intact ones. The amount of rice falling down the small chute into a tub is checked. Too many broken kernels indicate that the dehuller or screening machine requires adjustment.

The dehulled and scarified kernels are passed to a screening machine (Fig. 42), which has vibrating sieves that remove any foreign material such as stone and twigs, which may still remain. The rice then passes to a dimpler machine (Fig. 43), which separates broken from intact kernels. The broken kernels can be used in "redi-cooked" cans of wild rice or packaged in a mixture with white rice.

The large, unbroken, dark and polished kernels that are preferred in the culinary trade are finally sent to the gravity table (Fig. 44). In this final stage of processing, forced air causes the rice to separate out according to weight. There are no grading standards in Canada, but individual processors usually have their own grades. The heaviest and longest grain wild rice becomes their first-quality product, with the lighter, shorter grain having progressively lower grades. The graded wild rice is inspected and placed into 45-kg bags ready for bulk shipment or directly packaged into small display cartons or bags of 100, 250, or 500 g for the retail trade.

## Food value and uses

Wild rice is high in protein, carbohydrates, and minerals but low in fats (see Table 2). Like other cereal grains, wild rice does not contain vitamins A and C, but it is rich in the vitamin B complex, particularly riboflavin. The fat content of wild rice is less than 1%, but the high proportions of linolenic and linoleic fatty acids (up to 38%) make this small percentage of fat in wild rice very nutritional. Wild rice has a bland, nutty taste and is used in a wide variety of dishes, for example, dressings, soups, salads, and desserts (Fig. 45). Recently wild rice has been marketed in breakfast cereals and pancake, muffin, and cookie mixes. It qualifies as a natural food and is popular among health food enthusiasts because the watersheds of most natural commercial stands of wild rice are usually far removed from agrochemical residues.

HARVESTING AND PROCESSING THE GRAIN 93

Fig. 44. A gravity table separating wild rice grains into size classes.

Fig. 45. A wild rice casserole, a gourmet dish.

## Table 2. Nutrient composition of wild rice[1]

| Component | Composition |
|---|---|
| | *(% total dry weight)* |
| Moisture | 7.9 – 11.2 |
| Protein | 12.4 – 15.0 |
| Fat | 0.5 – 0.8 |
| Ash | 1.2 – 1.4 |
| Crude fiber | 0.6 – 1.1 |
| Total carbohydrate | 72.3 – 75.3 |
| Fatty acids | *(% total fatty acids)* |
|   Palmitic | 14 |
|   Stearic | 1 |
|   Oleic | 6 |
|   Linoleic | 8 |
|   Linolenic | 30 |
| Minerals | *(mg/100 g)* |
|   Calcium | 17 – 22 |
|   Magnesium | 80 – 161 |
|   Phosphorus | 298 – 400 |
|   Potassium | 55 – 344 |
|   Zinc | 3 – 6 |
|   Iron | 4 |
| Vitamins | *(mg/100 g)* |
|   Thiamine | 0.45 |
|   Riboflavin | 0.63 |
|   Niacin | 6.20 |

[1] Data from Anderson (1976).

# VII

# The Wild Rice Industry

## Government involvement

Statistics Canada attempts to keep records of wild rice exports, as shown in Table 3, but the numbers are inexact because there is no check on the values reported, and often there is a failure to differentiate between processed and unprocessed wild rice. Agriculture Canada gathers information on developments in the industry and is sometimes involved through small grants for research or through the New Crop Development program.

There is unrestricted trade in wild rice between Canada and the United States. Canadian exports of green or bulk processed wild rice are assessed a 2.5% ad valorem duty. A 10% ad valorem duty is levied on packaged or further refined wild rice products.

Some provincial governments are interested in promoting wild rice and they have regulations concerning the crop. In Quebec, New Brunswick, and Prince Edward Island, the ministries of agriculture are responsible for wild rice; in Nova Scotia, the Department of Lands and Forests of the Wild Life Division; in Ontario, the Ministry of Natural Resources; in Manitoba, both the ministries of Agriculture and Natural Resources; and in Saskatchewan, the Department of Parks and Renewable Resources are involved. In 1981, a research program was initiated at Lakehead University by the Department of Northern Affairs, Ontario, with the express aim of improving the farming of wild rice in natural lake systems.

Table 3. Exports of Canadian Wild Rice, 1978-1981[1]

| Country | Quantities (hundredweight) | | | | Value (thousands of dollars) | | | |
|---|---|---|---|---|---|---|---|---|
| | 1978 | 1979 | 1980 | 1981 | 1978 | 1979 | 1980 | 1981 |
| Australia | – | – | – | – | – | – | – | 10 |
| Belgium-Luxembourg | – | 5 | 19 | 3 | 4 | – | – | – |
| France | 5 | 13 | 20 | 20 | 3 | 9 | 25 | 15 |
| Germany | 14 | 34 | 57 | 100 | 8 | 20 | 39 | 65 |
| Italy | 2 | – | – | – | 1 | – | – | – |
| Sweden | – | 34 | – | – | – | 17 | – | – |
| Switzerland | 33 | 54 | 63 | 20 | 19 | 33 | 44 | 15 |
| United Arab Emirates | 394 | – | – | – | 10 | – | – | – |
| U.K. | 3 | 2 | – | – | 1 | 1 | – | – |
| U.S. | 6811 | 7258 | 14 905 | 11 540 | 1355 | 2736 | 1856 | 2687 |
| Venezuela | 12 | 24 | 15 | 7 | 22 | 14 | – | – |

[1] Data supplied by Statistics Canada 1983.

## *Alberta*

Permits to grow wild rice commercially are obtained from Alberta Energy and Natural Resources; each lake must pass through a review process before a permit is issued. By 1985, about 100 lakes were licensed.

Interest in growing wild rice in Alberta increased during the 1980s encouraged by the successes in Saskatchewan, and in 1983 the Northern Alberta Wild Rice Growers' Association (NAWRGA) was organized. This group works closely with Alberta Agriculture, which has jurisdiction over the wild rice crop produced in the province. In 1985, Agriculture Canada and the NAWRGA started a joint project to determine the potential for wild rice development in Alberta. An assessment of lakes for their soil type, presence of weeds, and general growth capacity for wild rice was begun. New, more efficient types of wild rice harvesters were also being designed. Although the wild rice industry in Alberta was only in the early early stages of development in 1985, the industry appears to have a bright future.

## Saskatchewan

Wild rice was introduced into northern Saskatchewan during the mid thirties. It was not until the late sixties and the early seventies that the potential of wild rice to contribute to the local economy of northern Saskatchewan was realized. The major organization concerned with wild rice in Saskatchewan is the Saskatchewan Indian Agriculture Program, Incorporated (SIAP). In 1978 the reported production was 27 470 kg and in 1983 the yield increased to 250 210 kg. In 1983, the average price for green rice was $3.30/kg and this new crop represented a significant income for the economy of the region. In 1983, the production of wild rice in Saskatchewan was estimated at about 12% of Canadian production, with almost all the crop being harvested by 81 producers operating 71 mechanical harvesters. In 1984 production is reported to have increased to 61–64% of the Canadian production, or over 50% of all lake (non-paddy) wild rice harvested in North America. The number of producers and newly seeded lakes has increased each year since 1978. In 1981, Saskatchewan Wild Rice Co-operative opened a processing plant with a capacity of 227 kg/h for producing processed rice.

The commercial interest in growing wild rice in northern Saskatchewan has been promoted by government assistance in supplying seed for the establishment of new wild rice stands and in the introduction of a permit system that favors northern residents. The success of the program can be seen in Figs. 46 and 47, which show the recent trend in production and permit holders.

## Manitoba

On 31 March 1984, the provincial government of Manitoba enacted a *Wild Rice Act* that embodied all legislation governing the propagation and harvesting of wild rice and the wild rice industry under a single law. Some of the major points in this legislation include the following items:
- a 10-year production license for producing areas that is assignable and transferable with ministerial approval
- an annual fee structure for production licenses based on a 3-year average production
- a development license that provides incentive for the development of new lakes as a prerequisite for a production license for those lakes
- a block license for community areas and the Whiteshell Provincial Park, with provision for the issuance of sublicenses

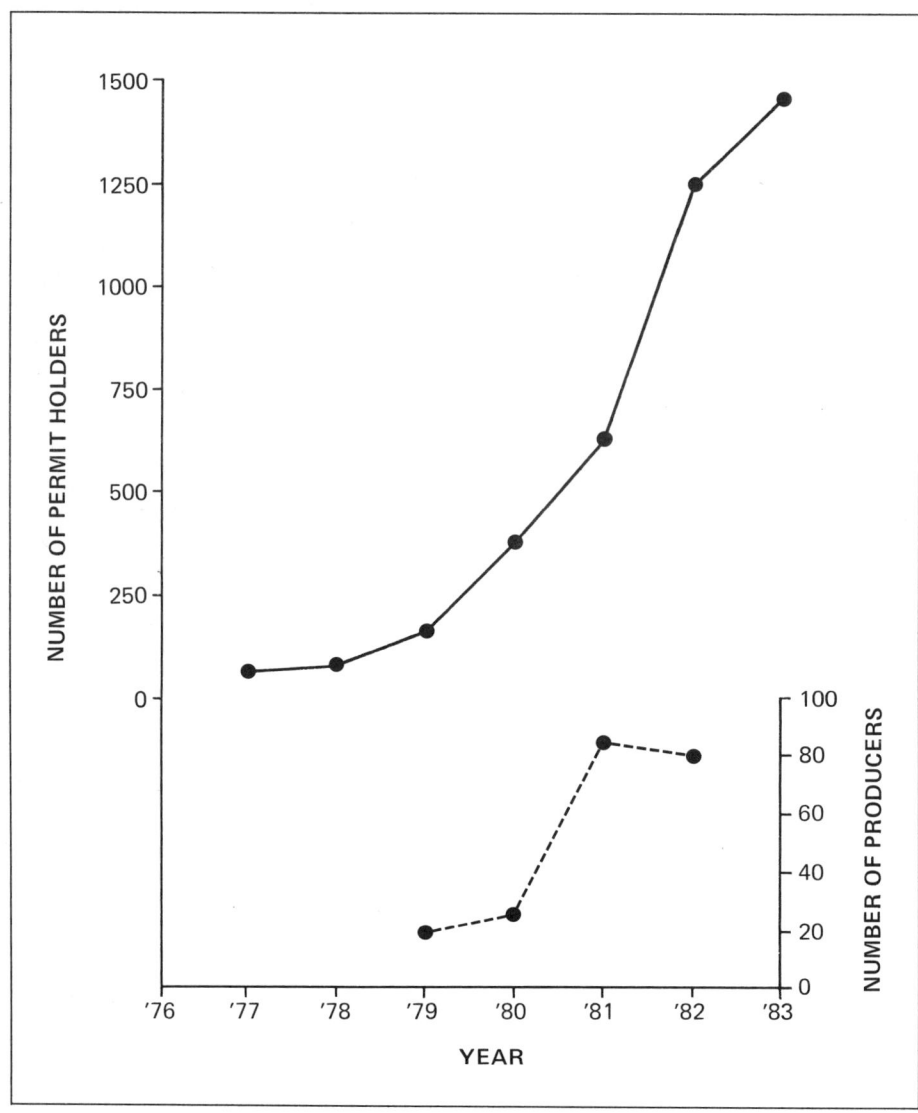

Fig. 46. Number of persons involved.

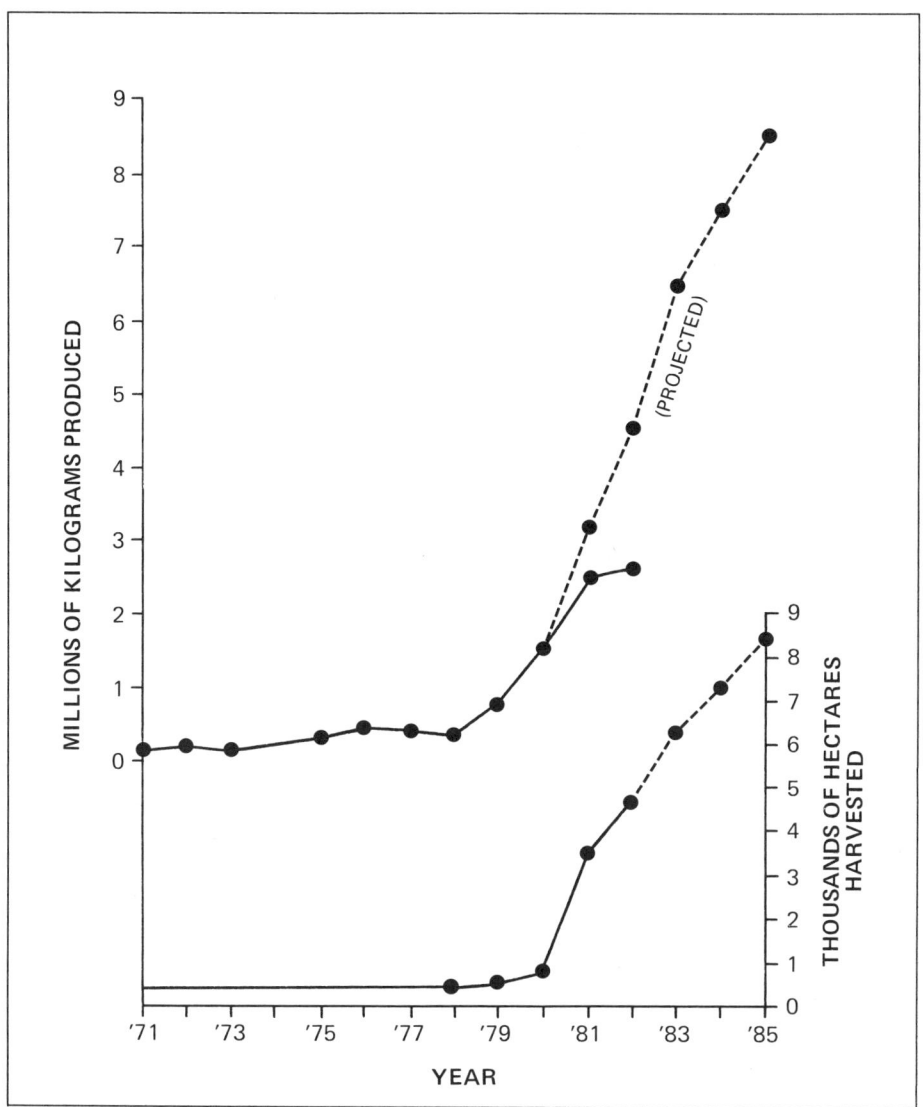

Fig. 47. Number of hectares and kilograms involved. (Source for statistics: Saskatchewan Indian Agricultural Program Inc.)

- the designation of zones and areas of the province to ensure the accommodation of local interests
- limits of allocation to any one person
- the introduction of a buyer's permit
- the introduction of a load slip
- a number of enforcement provisions that would facilitate the management of this renewable resource and the control of mechanical harvesters.

The Government of Manitoba proposes to license lakes and river areas to individuals, giving them the right to harvest any wild rice that grows in that particular area. The licensee is expected to enhance production by means of water control and the seeding of nonproductive sections of the licensed area.

The importance of wild rice to the provincial economy is given in Table 4, which shows the recorded production of unprocessed wild rice since 1970. The price structure multiplied by the recorded production gives an estimate of the monies paid at the harvester level. During 1979 a total of 143 licenses were issued, about half of them to native people. In 1986 more than 600 licenses went to 216 license holders and most of the expansion had taken place in the northern sector of the province. Both the native and non-native license holders continue to work with government and university researchers in developing management and harvest techniques in order to realize their wild rice production potential.

## *Ontario*

The framework for the harvesting regulations for wild rice in Ontario was provided by *The Wild Rice Harvesting Act* in 1960 as administered by the minister of Natural Resources, which provided for the following precedents in the management of wild rice on crown lands: wild rice is considered a provincial resource, harvesting privileges are reserved for residents of the province, and allowance is made for registration and licensing of harvesting areas on crown land.

The production of wild rice is entirely from natural water bodies, although considerable potential for paddy development exists. About 90% of the wild rice stands harvested commercially in Ontario are located in the northwestern region and the remainder is distributed throughout the north central and eastern regions.

Ontario's average production area of 8000 ha has varied dramatically over the past 7 years from 14 337 ha in 1977 to 3550 ha in 1982. One of the major factors affecting the total area is the annual

## Table 4. Manitoba Wild Rice[1]

| Year | Annual price structure, $/kg | Ten-year average, $/kg | Recorded production, kg |
|---|---|---|---|
| 1970 | 0.88 | – | 67 563 |
| 1971 | 1.10 | – | 226 720 |
| 1972 | 1.32 | – | 272 598 |
| 1973 | 1.54 | – | 284 489 |
| 1974 | 1.76 | – | 62 231 |
| 1975 | 1.98 | – | 64 622 |
| 1976 | 2.20 | – | 159 383 |
| 1977 | 2.87 | – | 523 614 |
| 1978 | 4.98 | – | 215 022 |
| 1979 | 4.70 | 2.34 | 271 058 |
| 1980 | 1.39 | 2.38 | 452 815 |
| 1981 | 3.44 | 2.62 | 205 396 |
| 1982 | 2.45 | 2.73 | 188 243 |
| 1983 | 3.26 | 2.91 | 152 255 |
| 1984 | 3.76 | 3.10 | 281 529 |
| 1985 | 2.31 | 3.14 | 293 098 |
| 1986 | 1.98 | 3.12 | 388 636 |
| 1987 | 1.23 | 2.95 | 752 273 |

[1] Data relating to unprocessed rice are provided by the Manitoba Department of Natural Resources. For 1970–1975, the price structure is estimated.

fluctuations of water levels, especially in the Lake of the Woods and Rainy Lake, where about half of the total wild rice stands are located. Annual production of wild rice varies widely, as seen from Table 5, where available records since 1963 are listed. Such oscillations in production are expected to continue until more refined culture techniques are practiced.

Since 1974 the development of new areas coupled with more intensive management and harvesting activities has increased significantly the economic status of wild rice in Ontario. Because of conflict within the wild rice industry, a 5-year moratorium was established in 1978 by a royal commission, which attempted to resolve the economic and social differences over the use of this resource. Though the moratorium has ended, negotiations between the Indian people and the province continue with the aim of protecting and enhancing native economic and social interests in this area as well as allowing expansion of the industry.

Table 5. Record of wild rice harvests (kg) from the Northwestern Region of Ontario

| Year | Dryden* | Fort Frances | Ignace | Kenora | Red Lake | Sioux Lookout | Northwestern Region | Av. price/kg ($) |
|---|---|---|---|---|---|---|---|---|
| 1963 | — | 1 608 | — | 21 352 | — | — | 22 960■ | 0.84 |
| 1964 | — | 719 | — | 22 904 | — | — | 23 623■ | 0.88 |
| 1965 | — | 2 394 | — | 10 134 | — | 101 | 12 629 | 1.14 |
| 1966 | — | 529 | — | 17 277 | — | 290 | 18 096 | 2.36 |
| 1967 | — | 16 647 | — | 207 045 | — | 9 044 | 232 736 | 2.53 |
| 1968 | — | 1 727 | — | 123 602 | — | 4 278 | 129 607 | 1.43 |
| 1969 | — | 11 473 | — | 53 173 | — | NA | 64 646■ | 1.04 |
| 1970 | — | 3 405 | — | 23 995 | — | NA | 27 401■ | 1.34 |
| 1971 | — | 13 804 | — | 95 260 | — | 15 830 | 124 894 | 1.23 |
| 1972 | — | 55 412 | — | 404 740 | — | 35 990 | 496 142 | 1.26 |
| 1973 | 27 840 | 27 876 | — | 511 660 | — | 18 053 | 585 429■ | 1.10 |
| 1974 | NA° | NA | — | 4 510 | — | NA | 4 510■ | 1.52 |
| 1975 | NA | 13 828 | — | NA | — | 28 514 | 42 342■ | 1.76 |
| 1976 | 33 176 | 64 651 | 3 918 | 343 732 | 13 702 | 57 382 | 516 561 | 1.56 |
| 1977 | 42 696 | 41 847 | 2 191 | 263 612 | 11 666 | 65 068 | 427 080 | 2.97 |
| 1978 | 26 738 | NA | 6 885 | 11 441 | 497 | 24 787 | 70 348■ | 4.40 |
| 1979 | 51 241 | 21 700 | 11 614 | 19 847 | 9 911 | 21 246 | 135 559 | 4.40 |
| 1980 | 75 174 | 54 864 | 16 775 | 197 938 | 26 042 | 69 239 | 440 032 | 1.65 |
| 1981 | 61 497 | 9 397■ | 17 264 | 125 041 | 24 841 | 72 080 | 310 120■ | 2.75 |
| 1982 | 3 525* | 454 | 7 856 | 12 697 | 2 541 | 58 435 | 85 508■ | 2.75 |
| 1983 | 11 565* | 826■ | 14 599 | 51 277 | 6 784 | 20 894 | 105 945■ | 2.50 |

* Incomplete data.
° NA, Not available.
■ The harvest data of Dryden, Ignace, and Red Lake before 1973 are included with the data of the old Kenora, Fort Frances, and Sioux Lookout administrative districts, respectively. Data are from *Wild Rice Report, 1983*, Ministry of Natural Resources, Northwestern Region, Kenora, Ont.

While development of this industry and expansion of the resource base have been delayed by the moratorium, several studies relating to factors that affect production have been completed. A research group based at Lakehead University and supported by the Department of Northern Affairs has studied potential improvements applicable to natural wild rice stands. With cooperation between all participants in the development of this resource, Ontario has the potential for significant increases in production of high-quality wild rice.

## New Brunswick

Wild rice has been planted in a number of lakes and rivers in New Brunswick (Steeves 1952), but it has not been harvested commercially. Ducks Unlimited is believed to do some planting for its own use. There is potential for wild rice production in selected areas, but the interest to date has been limited. There is no provincial legislation governing the harvesting of wild rice.

## Nova Scotia

In 1971, the Wildlife Division of the Department of Lands and Forests undertook to determine the feasibility of growing wild rice as a commercial crop in Nova Scotia. Wild rice is known to grow in large stands only in Long Lake on the LaPlanche River and Patton Lake on the Missiguash River near Amherst. The private interest in developing this natural resource in the province is minimal. There is no provincial legislation governing the harvesting of wild rice in natural stands.

## Prince Edward Island

Wild rice is harvested commercially at only one site, Pisquid Pond, where some 44–48 ha give yields up to 27 000 kg. There is no provincial legislation governing the harvesting of wild rice in this province.

## Other provinces

Wild rice is reported to have been planted in trials in Quebec, British Columbia, and the Yukon. The Department of Indian and Northern Affairs, in their newsletter *(Intercom* February–March 1985), reported an experimental seeding of wild rice in three lakes in the Yukon (Klukshy, Dezadeash, and Kloo) during October 1984. While the governments of Quebec and the Yukon have shown some interest in stimulating further industry since 1983, no commercial harvesting has yet been reported.

## Impact of paddy culture on the wild rice industry

Since the late sixties there has been a dramatic increase in the production of wild rice grown in paddies within the United States. This trend has continued so that today as much as 90% of the processed rice sold is from paddy culture. During the seventies Minnesota was the major producer, with smaller amounts grown in California, Florida, Idaho, Michigan, Washington, and Wisconsin. In the early eighties, a drop in the price of white rice grown in California led several farmers to try wild rice in their paddies. The result was so successful that in 1985 California replaced Minnesota as the leading state in wild rice production (Nelson and Dahl 1985).

Techniques used in paddy culture were developed in Minnesota and are described by Oelke et al. (1982). Average yields of unprocessed wild rice from paddies are reported to vary from 506 kg/ha in Minnesota to 1600 kg/ha in California. In contrast, yields from Canadian lakes rarely exceed 215 kg/ha. The higher yields from paddies are a function of advances in wild rice breeding that have resulted in the production of nonshattering strains, earlier-maturing varieties, and shorter plants; also, paddies are productive because they lend themselves to more efficient agronomic procedures (including fertilizer and pesticide applications), better control of water levels, and more efficient harvesting with modifie( white rice combines. Most research into enhanced production ( wild rice has taken place at the universities of Minnesota and Wisconsin and at the United States Department of Agriculture, Peoria, Ill.

In Minnesota, paddy expansion increased from 380 ha in 1968 to 7600 ha in 1973, and since 1979 the area cultivated has stabilized around 8500 ha. In 1983, Minnesota had 58 wild rice farms; the average area cultivated was 123 ha of paddies on each farm, with 12% of the farms having more than 422 ha of paddies. Of these large farms, 70% were family owned and produced about 60% of the total crop. No new farms were started between 1979 and 1983, partly as a result of new land use regulations. From 1974 production in Minnesota increased to 1 million kg of processed paddy rice in 1982. A goal of 2 million kg was planned for 1985, but only an estimated 1.8 million kg of processed wild rice was achieved. This figure represented an 11% increase over the 1984 harvest of 1.6 million kg.

The first recorded wild rice production in California was in 1977 when about 4500 kg of processed wild rice entered the market.

Areas assigned to wild rice production increased from 1000 ha in 1982, to 3200 ha in 1984, to over 6600 ha in 1985. Over 4 million kg of processed wild rice was produced during 1985, which was more than double the 1.7 million kg produced in 1984. Most of the wild rice is grown in the Sacramento Valley. There the climate is ideal and it is possible to harvest two crops of wild rice per year or to alternate wild rice with white rice.

## Organization of the wild rice industry

The six major stages in the marketing of wild rice involve the harvesters, buyers, processors, brokers, wholesalers, and retailers. The harvesters of natural stands sell their rice to lakeside buyers, who either purchase it on commission for a processor or wholesaler or act on their own to buy the grain and process it themselves. Most of the wild rice sales from processors are handled by brokers, who assemble specific orders and even package them for a commission. The price of processed grain depends on supply and demand, and prices have ranged from $4.40 to $12.20 per kilogram since 1968.

The Minnesota Wild Rice Growers Association, a grower cooperative, was formed in 1969 to lobby the government on behalf of the producers as well as to investigate production problems in the paddy culture of wild rice. In 1974 the growers voted to tax themselves 1 cent per pound on finished rice for product promotion. The tax has since been increased to 4 cents a pound and consideration is being given to a percentage tax rather than a fixed levy per pound. Another marketing cooperative of wild rice farmers known as United Wild Rice is the largest buyer in the United States. Some 60% of the wild rice crop is marketed through these two cooperatives; the remaining 40% is produced by independent operators, who often sell their product direct to the processor.

Much of the processed wild rice is bought by major companies such as Uncle Ben's and General Foods in the United States, who are believed to control up to 95% of the wild rice that is sold in mixes with white rice. The remainder of the crop is sold to restaurants, hotels, grocery chains, and specialty shops, and by private sales. Before 1967 most of the wild rice was sold in retail outlets and distributed in small packages. Although most Canadian wild rice is sold on the United States market, there has been a concentrated effort to diversify sales to other countries, notably to Europe and Japan (see Table 3).

The International Wild Rice Council was formed in 1975 to promote growth and development of the industry. This council is composed of producers, processors, and marketers from the United States and Canada, and it represents both paddy and lake interests. It attempts to increase sales by distributing promotional literature, such as recipes and articles of general interest, to newspapers, magazines, and journals. Wild rice from the natural stands is regarded as a gourmet food item; however, the production of large quantities of wild rice from paddy operations has made it necessary to expand the traditionally gourmet market.

## Industry quality control

The quality profile of wild rice can change drastically with variations in the processing procedures, but uniform quality standards have yet to be adopted. Many retailers judge quality on the unbroken state of the kernels, their color, and size. Broken kernels with their white centers detract from the appearance of the finished product. Variation in nutritive values has yet to be considered seriously as an aspect of quality. Many processors feel that color should be sacrificed to shorten the cooking time in response to the demand from commercial buyers for rapid cooking. Most processors also believe that consumers prefer a slightly toasted flavor.

## Outlook

The main goal of the industry has been to produce greater quantities of wild rice from both paddies and lakes. This goal was met between the early 1970s and 1984, when there was an average production increase of 42% a year (Wild Rice Growers Reports 1970–1986). Such increases were achieved by enlarging the total area cultivated or seeded and by increasing the productivity per hectare.

The total production of processed wild rice in both Canada and the United States has risen dramatically during the past 3 years, from 2.4 million kg in 1982 to 4 million kg in 1984, to over 6.4 million kg in 1985. In early 1986, the concern was that marketing procedures had not kept pace with the rapid increase in production. The establishment of an international marketing board and the expansion of sales in Europe, Australia, and Japan are currently being promoted.

When wild rice was harvested only from natural stands, it sold as a gourmet food. Canadian producers have attempted to present the longer-grain, lake-grown product as the gourmet version of the product grown in an environment free from agrochemical residues. There has been some lobbying to obtain legislation to identify the source where the rice is grown and its residue-free status on packages.

There have been attempts during the last 15 years to introduce paddy culture into Canada: in Manitoba at Brokenhead, Fort Alexander, Great Falls, Koostatak, Piney, Sprague, Starbuck, The Pas, and Whitemouth; and in northwestern Ontario at Dryden and Emo. But production from paddies in Canada is not yet significant. Canadians have preferred to exploit their less capital-intensive natural stands in lakes and rivers. Research in Canada has concentrated on water level control in natural systems, extensive seeding of wild rice from existing stands into areas where wild rice has not previously grown, the exploratory addition of fertilizers and weed control procedures to selected lakes, and the increased use of newer designs for mechanical harvesters. By applying research and development principles, Saskatchewan has increased production by over 500% between 1980 and 1984. Alberta also shows great promise in producing similar gains. Time, investment, cooperation, and interest are needed to ensure that wild rice production from lakes will continue to be economically feasible in Canada.

# References

Adams, J. B. 1945. Aphids on Canada wild rice. Can. Entomol. 77:196.

Aiken, S. G. 1986. The distinct morphology and germination of the grains of two species of wild rice *(Zizania,* Poaceae). Can. Field-Nat. 100(2):237–240.

Alex, J. F.; Cayouette, R.; Mulligan, G. A. 1980. Common and botanical names of weeds in Canada. Agric. Can. Publ. 1397. 132 pp.

Anderson, R. A. 1976. Wild rice: nutritional review. Cereal Chem. 53:949–955.

Anonymous. 1960. Index of plant diseases in the United States. U.S. Dep. Agric., Handb. No. 165. 531 pp.

Archibold, O. A.; Weichel, B. J. 1983. An ecological investigation of factors affecting the production of wild rice in northern Saskatchewan. Saskatchewan wild rice research. Sask. Agric. pp. 4–34.

Behan, M. J.; Kinraide, T. B.; Selser, W. I. 1979. Lead accumulation in aquatic plants from metallic sources including shot. J. Wildl. Manage. 43:240–244.

Berger, P. H.; Percich, J. A.; Ransom, J. K. 1981. Wheat streak mosaic virus in wild rice. Plant Dis. 65:695–696.

Boivin, B. 1967. Énumération des plantes du Canada. VI. Monopsides ($2^e$ partie). Nat. Can. (Que.) 94:471–528.

Bondar, M. I. 1958. Lake rice *(Zizania aquatica* – a prospective forage plant. Moskja. Akad. Nauk SSSR (In Russian). pp. 496–502.

Bowden, R. L.; Percich, J. A. 1983a. Bacterial brown spot of wild rice. Plant Dis. 67:941–943.

Bowden, R. L.; Percich, J. A. 1983b. Etiology of bacterial leaf streak of wild rice. Phytopathology 73:640–645.

Brooks, E. R. 1981. A wild rice program for the northwest Ontario Indian Reserves. Rep. northwestern Ont. Div., Dep. of Indian Affairs, Canada. 119 pp.

Brown, A. M. 1948. Ergot of cereals and grasses. Proc. Can. Phytopathol. Soc. (1947), 15:15.

Brown, E.; Scofield, C. S. 1903. Wild rice, its uses and propagation. U.S. Dep. Agric., Bur. Plant Ind., Bull. 50, 24 pp.

Brown, W. V. 1950. A cytological study of some Texas Gramineae. Bull. Torrey Bot. Club 77:63–76.

Cardwell, V. B.; Oelke, E. A.; Elliot, W. A. 1978. Seed dormancy mechanisms in wild rice *(Zizania aquatica)*. Agron. J. 70:481–488.

Chambliss, C. E. 1922. Wild rice. U.S. Dep. Agric. Circ. 229. 16 pp.

Chambliss, C. E. 1940. The botany and history of *Zizania aquatica* L. (wild rice). J. Wash. Acad. Sci. 30:185–205. (Reprinted in Smithsonian Inst. Annu. Rep. for 1940. pp. 369–382.)

Coltas, M. 1983. Biology of *Zizania* and preliminary investigations of the fruit production of the *Zizania* on Mud Lake. Unpublished B.Sc. Thesis, Carleton University. 79 pp.

Conners, I. L. 1939. 18th Annual Report, Can. Plant Dis. Surv. 1938. 112 pp.

Conners, I. L. 1967. An annotated index of plant diseases in Canada and fungi recorded on plants in Alaska, Canada and Greenland. Can. Dep. Agric. Publ. 1251. 381 pp.

Counts, R. L. 1983. Variation in the growth of Ontario wild rice. *In* The aquaculture of wild rice, progress year 2. Lakehead University.

Dale, J. E. 1979. Application equipment for roundup – the rope wick applicator. Proc. Beltwide Cotton Prod. Annu. Conf. 2 pp.

Darbyshire, S. J.; Aiken, S. G. 1986. *Zizania aquatica* var. *brevis* (Poaceae): a scanning electron microscope study of epidermal features. Nat. Can. (Que.) 113:355–360.

Davis, P. 1979. Impacts of commercial wild rice production on water quality in Minnesota. Rep. Minnesota Pollution Control Agency. pp. 100.

Densmore, F. 1974. How Indians use wild plants for food, medicine and crafts. Dover Pub, Inc., N.Y. pp. 279–397.

Dore, W. G. 1969. Wild rice. Ottawa: Agric. Can. Publ. 1393, 84 pp.

Dore, W. G.; McNeill, J. 1980. Grasses of Ontario. Agric. Can. Monogr. 26. 566 pp.

Ebsary, B. A.; Pharoah, G. 1982. *Hirschmaniella pisquidensis* n. sp. (Nematoda: Pratylenchidae) from roots of wild rice in Prince Edward Island, Canada. Can. J. Zool. 60:165–167.

Elliott, W. A. 1974. Wild rice breeding research. Prog. Rep. Wild Rice Res., Minnesota Agric. Exp. Stn., St. Paul. pp. 18–31.

Elliott, W. A. 1980. Wildrice. *In* Hybridization of crop plants. Crop Sci. Soc. Am.; Madison, Wis. pp. 721–731.

Emery, H. P. 1977. The current status of Texas wild rice, *Zizania texana*. Southwest Nat. 22:393–394.

Fassett, N. C. 1924. A study of the genus *Zizania*. Rhodora 26:153–160.

Fassett, N. C. 1927. *Zizania*. *In* notes from the herbarium of the Univ. Wisconsin. Rhodora 29:228–229.

Fernald, M. L. 1950. Gray's manual of botany. 8th ed. American Book Co., New York, N.Y. 1632 pp.

Ferren, W. R., Jr.; Good, R. E. 1977. Habitat, morphology and phenology of southern wild rice, *Zizania aquatica* from the Wading River in New Jersey. Bull. N.J. Acad. Sci. 104:392–396.

Fyles, F. 1915. A preliminary study of ergot of wild rice. Phytopathology 5:186–192.

Fyles, F. 1920. Wild rice. Agric. Can Bull. 42, 20 pp.

Gilbert, P. F. 1974. A preliminary study of fungi associated with diseases of wild rice in Manitoba. M.Sc. Thesis, University of Manitoba, 258 pp.

Gleason, H. A.; Cronquist, A. 1963. Manual of the vascular plants of the northeastern United States and adjacent Canada. Van Nostrand, Princeton, N.J. 810 pp.

Gronovius, J. F. 1743. Flora Virginica. Part 2. Lugduni Batavorum. pp. 129–206.

Hammond, G. H. 1958. A wild rice borer *(Apamea apamiformis* Gn.). Can. Insect Pest Rev. 36:254.

Hammond, G. H. 1959. Insect conditions in eastern Ontario, 1958. Can. Insect Pest Rev. 37:77.

Hawthorn, W.; Stewart, J. M. 1970. Epicuticular wax forms on leaf surfaces of *Zizania aquatica*. Can. J. Bot. 48:201–205.

Horner, D. 1983. Wild rice production in Saskatchewan. Sask. Indian Agriculture Program, Inc. 15 pp.

Jenks, A. E. 1901. The wild rice gatherers of the Upper Lakes. U.S. Dep. Int., Bur. Am. Ethnol., 9th Rep. (1899):1015–1160.

Johnson, E. 1969. Archeological evidence of the utilization of wild rice. Science 163:276–277.

Kardin, M. K.; Bowden, R. L.; Percich, J. A.; Nickelson, L. J. 1982. Zonate eyespot of wild rice caused by *Drechslera gigantea*. Plant Dis. 66:737–739.

Kernkamp, M. F.; Kroll, R.; Woodruff, W. C. 1976. Diseases of cultivated wild rice in Minnesota. Plant Dis. Rep. 60:771–775.

Lee, P. F. 1974. Methodology for the selection of sites for the artificial dissemination of wild rice in northwestern Ontario. Ont. Minist. Nat. Resour.

Lee, P. F. 1975a. Ecological factors in influencing the distribution and production of wild rice, *Zizania aquatica* L. in northwestern Ontario. Kenora:Ont. Minist. Nat. Resour. Rep.

Lee, P. F. 1975b. The feasibility of increasing the supply of wild rice, *Zizania aquatica* L., in northwestern Ontario by artificial seeding. Kenora:Ont. Minist. Nat. Resour. Rep.

Lee, P. F. 1979. Biological, chemical, and physical relationships of wild rice, *Zizania aquatica* L., in northwestern Ontario and northeastern Minnesota. Ph.D. Thesis, University of Manitoba, 174 pp.

Lee, P. F. 1982. The aquaculture of wild rice, progress year 1. Prepared for the Ontario Ministry of Northern Affairs, Lakehead University.

Lee, P. F. 1983a. Production of wild rice on a seeded lake near Ignace, Ontario. 34th Annu. AIBS Conf., University of North Dakota.

Lee, P. F. 1983b. The aquaculture of wild rice, progress year 2. Prepared for the Ontario Ministry of Northern Affairs, Lakehead University, 146 pp.

Lee, P. F. 1984. The acquaculture of wild rice, progress year 3. Rep. to Ministry of Northern Affairs, Ontario. 229 pp.

Lee, P. F.; Stewart, J. M. 1981. Ecological relationships of wild rice, *Zizania aquatica* L. 1. A model for among-site growth. Can. J. Bot. 59:2140–2151.

Lee, P. F.; Stewart, J. M. 1983. Ecological relationships of wild rice, *Zizania aquatica* L. 2. Sediment-plant tissue nutrient relationships. Can. J. Bot. 61:1775–1784.

Lee, P. F.; Stewart, J. M. 1984. Ecological relationships of wild rice, *Zizania aquatica* L. 3. Factors affecting seeding success. Can. J. Bot. 63:1608–1615.

Lips, E. 1956. Die Reisernte der Ojibwa-Indianer. Deut. Akad. Wiss. Akademi-Verlag. Berlin. 391 pp.

Lloyd, T. 1939. Wild rice in Canada. Can. Geogr. 19:288–299.

McAtee, W. L. 1917. Propagation of wild-duck foods. U.S. Dep. Agric., Bull. 465, 40 pp.

McCrimmon, H. R. 1968. Carp in Canada. Bull. Fish. Res. Board Can. 165, 93 pp.

MacKay, M. R.; Rockburne, F. W. 1958. Notes on life-history and larval description of *Apamea apamiformis* (Guenée), a pest of wild rice (Lepidoptera:Noctuidae). Can. Entomol. 90:579–582.

McQueen, D. A. R. 1981. A survey of the diseases of wild rice in Manitoba. M.Sc. Thesis, University of Manitoba. 90 pp.

Martin, A. C.; Uhler, F. M. 1939. Food of game ducks in the United States and Canada. U.S. Dep. Agric. Tech. Bull. 634, 147 pp.

Melanson, R. 1981. Lake fertilization in Nova Scotia, Black Duck Lake. N.S. Dep. Lands and Forests. 4 pp.

Melvin, J. C. E. 1966. Observations on insects attacking wild rice in Manitoba. Proc. Entomol. Soc. Manit. 22:6–11.

Moulton, D. W.; Weller, M. W. 1978. Studies in the control of blackbird damage to cultivated wild rice. Univ. Minn. Agric. Exp. Stn. Prog. Rep. 1977 Wild Rice Res. pp. 75–92.

Moyle, J. B. 1944. Wild rice in Minnesota. J. Wildl. Manage. 8:177–184.

Moyle, J. B. 1945. Some chemical factors influencing the distribution of aquatic plants in Minnesota. Am. Mid. Nat. 34:402–420.

Nelson, R.; Dahl, R. 1986. Wild rice market shows vigorous growth. *In* Minn. Wild rice research 1985. Misc. Publ. 36. Agric. Exp. Stn., Univ. Minn. pp. 70–75.

Nickels, N. 1952. Is the wild rice crop doomed? Rod and Gun 54:11,33.

Oelke, E. A.; Grava, J.; Noetzel, D.; Barron, D.; Percich, J.; Schertz, C.; Strait, J.; Stucker, D. 1982. Wild rice production in Minnesota. Univ. Minn. Ext. Bull. 464, 40 pp.

Orcajada, P. G. 1982. Growing wild rice in northern Saskatchewan. Dep. of northern Saskatchewan Rep. 9 pp.

Pantidou, M. E. 1959. *Claviceps* from *Zizania*. Can. J. Bot. 37:1233–1236.

Peden, D. G. 1977. Waterfowl use of exotic wild rice habitat in northern Saskatchewan. Can. Field-Nat. 91:286-287.
Percich, J. A.; Huot, K.; Kohls, C.; Schickli, L. 1985. Wild rice disease. Univ. Minn. Agric. Exp. Stn., Minn. Wild Rice Res. 1984, pp. 51-73.
Percich, J. A.; Nickelson, L. J.; Bowden, R. L.; Kardin, M. K. 1981. Wild rice disease research - 1980. Univ. Minn. Agric. Exp. Stn., Minn. Wild Rice Res. 1980. pp. 37-47.
Peterson, A. G.; Noetzel, D. M.; Sargent, J. E.; Hanson, P. E.; Johnson, C. B.; Soemawinata, A. T. 1981. Insects of wild rice in Minnesota. Univ. Minn. Agric. Exp. Stn., Misc. Rep. 157, 15 pp.
Powles, P. M.; MacCrimmon, H. R.; Macrae, D. A. 1983. Seasonal feeding of carp, *Cyprinus carpio,* in the Bay of Quinte watershed, Ontario. Can. Field-Nat. 97:293-298.
Punter, D.; Reid, J.; Hopkin, A. A. 1984. Notes on sclerotium-forming fungi from *Zizania aquatica* (wildrice) and other hosts. Mycologia 76:722-732.
Richards, W. R. 1960. A synopsis of the genus *Rhopalosiphum* in Canada (Homoptera:Aphididae). Can. Entomol. Suppl. 13, 51 pp.
Robinson, A. G.; Bradley, G. A. 1968. A revised list of the aphids of Manitoba. Manit. Entomol. 2:60-65.
Sain, P. 1983. Decomposition of wild rice *(Zizania aquatica* L.) straw and its effect on the depletion of oxygen during winter in natural lakes of northwestern Ontario. Ont. Fish. Tech. Rep. Ser. 8. 4 pp.
Sanwal, K. C. 1957. The morphology of the nematode *Radopholus gracilis* (de Man, 1880) Hirschmann, 1955, parasitic in roots of wild rice, *Zizania aquatica* L. Can. J. Zool. 35:75-92.
Sargent, J. E. 1976. Biology of the wild rice stalk borer. Univ. Minn. Agric. Exp. Stn., Prog. Rep. of 1975 Wild Rice Res. pp. 35-37.
Sculthorpe, C. D. 1967. The biology of aquatic vascular plants. Edward Arnold, London. 610 pp.
Simpson, G. M. 1966. A study of germination in the seed of wild rice *(Zizania aquatica).* Can. J. Bot. 44:1-9.
Steeves, T. A. 1952. Wild rice - Indian food and a modern delicacy. Econ. bot. 6:107-142.
Svare, C. W. 1960. The effects of various oxygen levels on germination and early development of wild rice. Minn. Dep. Conserv., Div. Game Fish, Game Invest. Rep. 3.
Taber, W. A.; Vining, L. C. 1960. A comparison of isolates of *Claviceps* spp. for the ability to grow and to produce ergot alkaloids on certain nutrients. Can. J. Microbiol. 6:355-365.
Terrell, E. E.; Batra, S. W. T. 1982. *Zizania latifolia* and *Ustilago esculenta,* a grass-fungus association. Econ. Bot. 36:274-285.
Terrell, E. E.; Batra, S. W. T. 1984. Insects collect pollen on eastern wildrice, *Zizania aquatica* (Poaceae). Castanea 49:31-34.
Terrell, E. E.; Emery, W. H. P.; Beaty, H. E. 1978. Observations on *Zizania texana* (Texas wildrice), an endangered species. Bull. Torrey Bot. Club 105:50-57.

Terrell, E. E.; Wergin, W. P. 1979. Scanning electron microscopy and energy dispersive X-ray analysis of leaf epidermis in *Zizania* (Gramineae). Scanning Electron Microsc. III:81–88.

Terrell, E. E.; Wergin, W. P. 1981. Epidermal features and silica deposition in lemmas and awns of *Zizania* (Gramineae). Am. J. Bot. 68:697–707.

Thieret, J. W. 1971. Observations on some aquatic plants in northwestern Minnesota. Mich. Bot. 10:117–118.

Thomas, A. G.; Stewart, J. M. 1969. The effect of different water depths on the growth of wild rice. Can. J. Bot. 47:1525–1531.

Warwick, S. E.; Aiken, S. G. 1986. Electrophoretic evidence for the recognition of two species in annual wild rice *(Zizania,* Poaceae). Syst. Bot. 11:464–473.

Weber, R. P.; Simpson, G. M. 1967. Influence of water on wild rice *(Zizania aquatica* L.) grown in a prairie soil. Can. J. Plant Sci. 47:657–663.

Wiegand, K. M.; Eames, A. J. 1925. Flora of the Cayuga Lake Basin, New York. Cornell Univ. 491 pp.

Wild Rice Growers Reports. 1970–1986. Obtainable from the International Wild Rice Assoc., P.O. Box 366, Aitkin, Minn. 56431.

Woods, D.; Gutek, L. 1974. Germinating wild rice. Can. J. Plant Sci. 54:423–424.

Wright, M. C. 1942. An investigation of ergot on wild rice. M.Sc. Thesis, University of Maine.

# Additional Reading

Anderson, R. A. 1978. Wild rice: its history, current production, use. Rice J. 81:34–38.
Anderson, R. A.; Navickis, L. L.; Warner, K. A.; Vojnovich, C.; Bagley, E. B. 1979. Quality characteristics of processed wild rice *(Zizania aquatica)*. Cereal Chem. 56:375–379.
Anderson, R. A.; Vojnovich, C.; Navickis, L. L.; Bagley, E. B. 1979. Parching studies on wild rice *(Zizania aquatica)*. Cereal Chem. 56:371–374.
Atkins, T. A. 1983. The aquaculture of wild rice, progress year 2. Addendum. An investigation of the seasonal trends and relationships of wild rice growth and its physio-chemical environment at Lake of the Woods. Report to Ontario Ministry of Northern Affairs, Lakehead University, Thunder Bay, Ont. 50 pp.
Baldwin, W. W. 1958. Plants of the Clay Belt of northern Ontario and Quebec. Natl. Mus. Can. Bull. 156 pp.
Bean, G. A.; Schwartz, R. 1961. A severe epidemic of *Helminthosporium* brown spot disease on cult. wild rice in Minnesota. Plant Dis. Rep. 45:901.
Bean, W. J. 1909. The Canadian wild rice. Kew Bull. 9:380–385.
Belitzer, N. V. 1963. On the embryology of *Zizania aquatica* (in Russian). Bot. Zh. 20:7–15.
Borys, A. 1980. Wild rice, Manitoba's most historic crop. Conservation Comment. Dep. Natural Resources, Government of Manitoba. 8 pp.
Brogan, D. 1956. Wild rice harvest. Frontiers of Plant Sci. pp. 131–135.
Campiranon, S.; Koukkari, W. L. 1977. Germination of wild rice *Zizania aquatica* seeds and the activity of alcohol dehydrogenase in young seedlings. Physiol. Plant. 41:293–297.
Camus, A. 1950. Les espèces utiles du genre *Zizania*. Rev. Int. Bot. Appl. Agric. Trop. 30:50–62.
Capen, R. G.; Leclerc, J. A. 1948. Wild rice and its chemical composition. J. Agric. Res. 77:65–79.

Chan, Y. S.; Thrower, L. B. 1980. The host parasite relationship between *Zizania carduciflora* and *Ustilago esculenta:* 1. Structure and development of the host and host parasite combination. 2. *Ustilago esculenta* in culture. 3. Carbohydrate metabolism of *Ustilago esculenta* and the host parasite combination. New Phytol. 85:201–234.

Chang, H. S. 1974. Intercross fertility between *Helminthosporium oryzae*, *Helminthosporium zizaniae*, and an unidentified *Helminthosporium* sp. on *Zizania aquatica*. Bot. Bull. Acad. Sin. (Jaipaei) 15:103–111.

Chang, H. S. 1977. Light inhibits sclerotial formation of an isolate of *Helminthosporium sigmoideum* which causes stem rot in *Zizania latifolia*. Bot. Bull. Acad. Sin. (Jaipei) 18:39–44.

Cooper, L. R. 1953. Wild rice gathering and processing. Minneapolis: Minn. Nat. 3:57–60.

Dale, H. M.; Miller, G. E. 1978. Changes in the aquatic macrophyte flora of Whitewater Lake near Sudbury, Ont., Canada, from 1947–1977. Can. Field-Nat. 92:264–270.

Davids, R. 1935. The legend of the wild rice (Chippewa). Univ. Minn.: Literary Rev. Vol. II (No. 1), 3rd quarter, 2 pp.

DeWet, J. M. J.; Oelke, E. A. 1978. Domestication of American wild rice *(Zizania aquatica* L. Gramineae). J. Agric. Trop. Bot. Appl. 25:67–84.

Duvel, J. W. T. 1906. The germination and storage of wild rice seed. U.S. Dep. of Agric., Bur. Plant Ind. Bull 90:1–13.

Elliott, W. A.; Perlinger, G. J. 1977. Inheritance of shattering in wild rice. Crop Sci. 17:851–853.

Everett, L. A.; Stucker, R. E. 1983. A comparison of selection methods for reducing shattering in wild rice. Crop Sci. 23:956–960.

Foster, K. W.; Rutger, J. N. 1980. Genetic variation of 4 traits in a population of *Zizania aquatica*. Can. J. Plant Sci. 60:1–4.

Goel, M. C.; Marth, E. H.; Stuiber, D. A.; Lund, D. B.; Lindsay, R. C. 1972. Changes in the micro flora of wild rice during curing by fermentation. J. Milk Food Technol. 35:385–391.

Goto, K.; Fukazu, R.; Chata, K. 1953. Overwintering of the causal bacteria of rice leaf blight in the rice plant and grasses. Agric. and Hort. 28:207–208.

Grava, J.; Raisanen, K. A. 1978. Growth and nutrient accumulation and distribution in wild rice. Agron. J. 70:1077–1081.

Gutek, L. H.; Woods, D. L.; Clark, K. W. 1981. Identification and inheritance of pigments in wild rice *(Zizania aquatica)*. Crop Sci. 21:79–82.

Hallowell, A. I. 1935. Notes on the northern range of *Zizania* in Manitoba. Rhodora 37:302–304.

Halstead, E. H.; Vicario, B. T. 1969. Effect of ultra sonics on the germination of wild rice *(Zizania aquatica)*. Can. J. Bot. 47:1638–1640.

Hance, H. F. 1872. On a Chinese culinary vegetable *(Zizania latifolia)*. J. Bot. Brit. Foreign 10:146–149.

Hanten, H. B.; Ahlgren, G. E.; Carlson, J. B. 1980. The morphology of grain abscission in *Zizania aquatica*. Can. J. Bot. 58:2269–2273.

Hirayoshi, I. 1956. Chromosomal relationships in Oryzae and Zizanieae. Proc. Int. Genet. Symp. Tokyo Kyoto, pp. 293–297.

Hofstrand, R. H. 1970. Wild ricing. Nat. Hist. 79:50–55.

Hong-ji, Su. 1975. Some studies on the cytological cultural characters and cultural variants of *Ustilago esculenta*. (*Zizania aquatica* galls). Kuo Li Tai-wan Hsueh Chin Wu Ping Ch'ung Hai Hsueh K'an 4:107.

Jarvenpa, R. 1971. Political entrenchment in an Ojibwa wild rice economy. J. Minn. Acad. Sci. 37:66–71.

Kaye, B.; Moodie, D. W. 1978. The Psoralea food resource of the Northern Plains. Plains Anthropol. 23:329–336.

Kaverzneva, I. G. 1960. The aerenchyma of *Zizania aquatica* (In Russian). Bot. Zh. Lennigr. 45:572–577.

Kernkamp, M. F.; Kroll, R.; Woodruff, W. C. 1977. Wild rice infected by *Sclerotium* sp. isolated from white water lily. Plant Dis. Rep. 61:187–188.

Kim, J. M.; Lorenz, K. 1981. Enzymatic activities in wild rice *(Zizania aquatica)*. Lebensm.-Wiss. Technol. 14:23–27.

Lee, P. F. 1976. The incredible potential of wild rice. Ont. Fish Wildl. Rev. 15:19–20.

Lee, P. F. 1986. Summary report: the aquaculture of wild rice. Lakehead University. 42 pp.

Lin, C. H.; Chang, L. R. 1978. Anatomical approach of the crown gall formation of *Zizania aquatica*. Plant Physiol. (Bethesda) 61:73.

Lindenfelser, L. A.; Ciegler, A.; Hesseltine, C. W. 1978. Wild rice as fermentation substrate for myco toxin production. Appl. Environ. Microbiol. 35:105–108.

Lorenz, K. 1981. The starch of wild rice *(Zizania aquatica)*. Starch Staerke 33:73–76.

Lund, D.; Lindsay, R.; Stuiber, D.; Johnson, C. E.; Marth, E. H. 1975. Drying and hulling characteristics of wild rice. Cereal Foods World 20:150–154.

Macins, V. 1969. Observations on the relation of water levels in Lake of the Woods to the wild rice crop in Kenora district. Kenora, Ont.: Ont. Minist. Nat. Resour. Rep. 12 pp.

Mackenzie, A. 1802. Voyage from Montreal on the River St. Lawrence through the continent of North America to the frozen and pacific oceans in the years 1789 and 1793. 2 Vol. London: R. Noble.

McAllister, D. 1976. Wild rice – old crop with new impact. Minn. Sci. 34:3–5.

McAndrews, J. H. 1969. Paleobotany of a wild rice lake in Minnesota. Can. J. Bot. 47:1671–1679.

Melson, J. W.; Podmer, L. S. 1942. The thiamine, riboflavin, nicotinic acid and pantothenic acid contents of wild rice *(Zizania aquatica)*. Cereal Chem. 19:539–540.

Melvin, J. C. E. 1960. Observations on insects attacking wild rice in the Whiteshell Forest Reserve. Winnipeg: Can. Dep. Agric., For. Biol. Div. Interim Rep. for 1959. 5 pp.

Melvin, J. C. E. 1962. Insects attacking wild rice in Manitoba. Can. Dep. For., Bi-mon. Progress Rept., 18:2 pp.

Miller, H. J. 1943. Wild rice in Michigan. Mich. Conserv. 12:4–5.
Morrison, R. H.; King, T. H. 1971. Stem rot of wild rice in Minnesota. Plant Dis. Rep. 55:498–500.
Moyle, J. B. 1941. Minnesota's wild rice crop. Conserv. Volunteer 20:30–37.
Moyle, J. B. 1942. The 1941 Minnesota wild rice crop. Dep. of Conservation, Div. Game Fish, Bur. Fish. Res. Invest. Rep. 40.
Moyle, J. B. 1945. Manomin – Minnesota's native cereal. Conserv. Volunteer 8:29–31.
Moyle, J. B. 1957. Minnesota's famous wild rice. Conserv. Volunteer 20:38–44.
Moyle, J. B. 1977. Wild rice: native grain of northern waters. J. Freshwater Biol. 15:18–21.
Moyle, J. B.; Krueger, P. 1964. Wild rice in Minnesota. Minnesota Conserv. Volunteer 27:30–37.
Navickis, L. L.; Anderson, R. A. 1978. Composition of a by-product of wild rice processing. Cereal Chem. 55:544–546.
Nisikado, Y. 1929. Studies on the *Helminthosporium* diseases of Gramineae in Japan. Ber. Ohara Inst. Landivirtsch Biol. Okayana Univ. 4:111–126, 11 plates.
Oelke, E. A. 1976. Amino-acid content in wild rice *Zizania aquatica* Grain. Agron. J. 68:146–148.
Oelke, E. A.; Albrecht, K. A. 1978. Mechanical scarification of dormant wild rice seed. Agron. J. 70:691–694.
Oelke, E. A.; Albrecht, K. A. 1980. Influence of chemical seed treatments on germination of dormant wild rice *(Zizania palustris)* seeds. Crop Sci. 20:595–598.
Oelke, E. A.; Brun, W. A. 1978. Paddy production of wild rice. Fact Sheet 20. Univ. Minn. Agric. Ext. Serv.
Oelke, E. A.; Elliot, W. A. 1978. Seeding time, method, and rate for wild rice grown as a field crop. Agronomy Fact Sheet 33. University of Minnesota. Agric. Ext. Serv.
Ogan, M. T. 1979. Potential for nitrogen fixation in the rhizosphere and habitat of natural stands of the wild rice *(Zizania aquatica)*. Can. J. Bot. 57:1285–1291.
Peterson, A. G.; Johnson, C. B. 1975. Observations on cultural control of the rice stalk in Minnesota. Proc. North Cent. Br. Entomol. Soc. Am. 30:81–82.
Plansearch, Inc. 1980. Feasibility of wild rice production in Nova Scotia. Rep. for Nova Scotia Dep. Agric. and Marketing.
Reagan, A. B. 1919. Wild or Indian rice. Proc. Ind. Acad. Sci. pp. 241–242.
Rogalsky, J. R.; Clark, K. W.; Stewart, J. M. 1971. Wild rice paddy production in Manitoba. Soils and Crops Branch. Manit. Dep. Agric. Publ. 527. 22 pp.
Sain, P. 1984. Decomposition of wild rice *(Zizania aquatica)* straw in two natural lakes of northwestern Ontario. Can. J. Bot. 62:1352–1356.
Schertz, C. E.; Boedicker, J. J.; Chinsuwan, W. 1980. Equipment and procedures for combine separation studies on wild rice. Trans. Am. Soc. Agric. pp. 309–311.

Schertz, C. E.; Oelke, E. A.; Skoe, R. C.; Barron, D. D. 1977. Wild rice harvest – new challenge for grain combines. Grain Forage Harvest. pp. 90–98.

Scofield, C. S. 1905. The salt water limits of wild rice. U.S. Dep. Agric., Bur. Plant Ind., Bull. 72. 8 pp.

Sinner, G. T.; Schramm, L. C. 1968. Preliminary phytochemical investigation of wild rice *(Zizania aquatica)* ergot. J. Pharmacol. Sci. 57:889–890.

Steeves, T. A.; DeWolf, G. P. 1950. A note on the varieties of *Zizania aquatica* L. Rhodora 52:34.

Stewart, J. M. 1970. Paddy production of wild rice on Muskegs. Proc. 13th Muskeg Res. Conf. NRC Tech. Mem. 94:91–97.

Stickney, G. P. 1896. Indian use of wild rice. Am. Anthropol. 9:115–121.

Stoddard, C. H. 1957. Utilization of swamplands for wild rice production: a progress report. J. Soil Water Conserv. 33:135–138.

Stoddard, C. H. 1960. Wild rice production from new wetlands. Trans. 25th North Am. Wildl. Conf. pp. 144–153.

Stone, G.; Stewart, J. M.; Woods, D.; Punter, D.; Beaubier, G. 1975. Wild rice production in Manitoba. Manit. Dep. Agric. Publ. 527. 2nd ed.

Stover, E. L. 1928. The roots of wild rice, *Zizania aquatica* L. Ohio J. Sci. 28:43–49.

Stuiber, D. A.; Johnson, C. E.; Lund, D. B.; Lindsay, R. C.; Brickbauer, E. A. 1972. Wild rice processor's handbook 1972. Cooperative Extension Program. University of Wisconsin Extension and Upper Great Lakes Regional Commission Publ., University of Wisconsin, Madison.

Suffling, R.; Schreiner, C. 1979. A bibliography of wild rice including biological, anthropological and socio-economic aspects. University of Waterloo, School of Urban and Regional Planning. Working Paper 5. 54 pp.

Swain, E. W.; Wang, H. L.; Hesseltine, C. W. 1978. Note on vitamins and minerals of wild rice. Cereal Chem. 55:412–414.

Taube, E. 1951. Wild rice. Sci. Mon. 73:369–375.

Terrell, E. E.; Wiser, W. J. 1975. Protein and lysine contents in grains of 3 species of wild rice. Bot. Gaz. 136:312–316.

Tsuda, M.; Ueyama, A. 1975. Identity of *Helminthosporium* leaf spot fungi attacking Oryziodeae plants growing in Japanese Islands. Prelim. Note. Nippon Kakingakri Kaishi 16:93–94.

Vaquer, A. 1973. Absorption and accumulation of pesticides residues and chlorinated biphenyls in both wild aquatic vegetation and rice in the Camargue Region. Oecol. Plant 4:353–365.

Want, H. L.; Swain, E. W.; Hesseltine; C. W.; Gumbmann, M. R. 1978. Protein quality of wild rice. J. Agric. Food Chem. 26:309–312.

Watts, B. M. 1980. Chemical and physicochemical studies of wild rice. Ph.D. Thesis, University of Manitoba, 155 pp.

Weber, R. P. 1967. A study of the feasibility of domesticating wild rice *(Zizania aquatica* L.). M.Sc. Thesis, Department of Crop Science, University of Saskatchewan, Saskatoon, Sask.

Weir, C. E.; Dale, H. M. 1960. Developmental study of wild rice, *Zizania aquatica* L. Can. J. Bot. 38:719-739.

Whigham, D.; Simpson, R. 1977. Growth, mortality, and biomass partitioning in freshwater tidal wetland populations of wild rice *(Zizania aquatica var. aquatica)*. Bull. Torrey Bot. Club 64:347-351.

Wirakartakusumah, M. A.; Lund, D. B. 1978. Kernel hardness of wild rice as affected by drying air temperatures and moisture gradient. J. Food Sci. 43:394-396.

Withycombe, D. A.; Lindsay, R. C.; Stuiber, D. A. 1978. Isolation and identification of volatile components from wild rice grain *(Zizania aquatica)*. J. Agric. Food Chem. 26: 816-822.

Woods, D. L.; Clark, K. W. 1976. Preliminary observations on the inheritance of nonshattering habit in wild rice. Can. J. Plant Sci. 56.

Yarnell, R. A. 1964. Aboriginal relationships between culture and plant life in the Upper Great Lakes Region, Anthropol. Pap. Mus. Anthropol. Univ. Mich. 23. University of Michigan, Ann Arbor, Mich.

# Glossary

**ad valorem**   A tax applied according to the value of the product.
**adventitious root**   A root that arises from any organ other than the primary root or branch.
**alkaloid**   An organic substance having alkaline properties: an organic base.
**anaerobic**   Applied to situations without oxygen, or to cells that can live in an environment without oxygen.
**anthesis**   The period when the flower is open and functional.
**ascospores**   Spores, usually eight, produced in an ascus, a typically sac-like cell.
**biomass**   Total dry weight of all organisms in a particular habitat or area.
**braconid parasite**   A member of the Braconidae, a group of parasitic insects that are usually less than 15 mm long.
**Bravo**   A commercial fungicide containing the active ingredient chlorothalanil.
**capitate**   Forming a head.
**caryopsis**   A simple, dry, indehiscent fruit, with pericarp firmly united all around the seed coat.
**cauline**   Belonging to or growing on a stem.
**cervical shield**   A stiff (chitinous) plate on some caterpillars, just behind the head.
**chlorosis**   The process leading to the destruction of green pigments in plants and resulting in a yellowish or whitish appearance.
**chlorotic**   An absence of chlorophyll giving a yellow or whitish appearance.

| | |
|---|---|
| **coalesce** | Growing together. |
| **coleoptile** | The cylindrical sheath protecting the embryonic shoot in a seed or seedling. |
| **conidium** | (plural conidia) A specialized, non-motile, asexual spore. |
| **cuticle** | A waxy or fatty layer on the outer wall of the epidermal cells. |
| **decumbent** | Growing upwards from a base that lies along the ground. |
| **dehulling** | Removing the hull, the flower scales around a seed. |
| **diptera(n)** | A member of the Diptera, a group of insects that have only one pair of wings, for example, deer flies, black flies, and mosquitoes. |
| **Dithane M45** | A fungicide containing the active ingredient mancozeb. |
| **ecocline** | The tendency for variation to follow an ecological gradient. |
| **ecotype** | Locally adapted variant of an organism. |
| **endemic** | Native in a restricted locality. |
| **endosperm** | Nutritive tissue in a grain, arising in the embryo sac of angiosperms following the fertilization of the two fused polar nuclei by a male gamete. |
| **epiblast** | A small nonvasculated flap of tissue between a grass embryo and the grain wall. |
| **epicuticular** | On the cuticle. |
| **epidemiology** | 1. A science that deals with the incidence, distribution, and control of disease in a population. 2. The sum of the factors controlling the presence or absence of a disease or pathogen. |
| **epiphyte** | An organism that grows upon another plant but is not parasitic upon it. |
| **eutrophic** | Waters rich in dissolved nutrients but frequently shallow with seasonal oxygen deficiency. |
| **geniculate** | Abruptly bent, as at the elbow or knee joint. |
| **genotype** | The genetic constitution of an individual. |
| **glaucescence** | Covered with a whitened or waxy bloom that wears off with age. |
| **glumes** | The pair of bracts present at the base of the male spikelet. |
| **hull** | The two interlocking spikelet scales that surround the caryopsis. |
| **ichneumonid parasite** | A member of the Ichneumonidae, a very large group of wasps, the majority of which resemble slender moths. |

| | |
|---|---|
| imperfect state (of fungi) | That portion of the life cycle characterized by asexual spores (conidia) or the absence of spores. |
| inoculation experiments | The communication of an infective agent to a healthy individual. |
| inoculum | Material used in or suitable for use in inoculation or inoculating. |
| instar | The period or stage between molts in the larva, for example, the first instar is the stage between the egg and first molt. |
| integument | Outermost layer or layers of tissue enveloping the nucellus of the ovule; develops into the seed coat. |
| intertidal | The zone between high tide and low tide. |
| isoenzyme | A multiple molecular form of an enzyme with similar or identical substrate specificity. It is coded by different gene loci. |
| larva(e) | A young insect that leaves the egg in an early stage of morphological development and looks different from the adult. |
| lemma | The lower of two bracts surrounding a grass flower. |
| lesion | Injury, impairment, flaw. |
| ligule | A membranous or hairy appendage on the adaxial surface of the grass leaf at the junction of sheath and blade. |
| mesocotyl | The elongating axis below the coleoptile in a seedling. |
| necrosis | The death of living tissue. |
| necrotic | Producing necrosis. |
| noctuid moth | A member of the Noctuidae, night-flying moths, including the majority of moths that are attracted to lights at night. |
| nomenclature | A system or set of names or designations. |
| palea | The upper of two bracts surrounding a grass flower. |
| panicle | When applied to grasses, an inflorescence in which all the spikelets have pedicels (short stalks). |
| paniculate | Having a panicle. |
| pathogen | A specific cause of disease (such as a bacterium or virus). |
| pedicel | The stalk of a single flower. |
| pedicellate | Having or attached to a slender stalk. |
| pericarp | Grain wall that is developed from the ovary wall. |
| perithecium(a) | Sexual fruiting body of Ascomycete fungi. |
| phenotype | The visible characteristics of an organism. |
| photosynthetic | Using, relating to, or formed by photosynthesis, the process in plants where water and carbon dioxide in the presence of chlorophyll and sunlight are converted into carbohydrates. |

| | |
|---|---|
| pith | The ground tissue occupying the center of the stem or root within the vascular cylinder. |
| pseudo-septum (a) | A protoplasmic or vacuolar membrane that looks like a wall. |
| pupation | The act of becoming a pupa, the resting, inactive stage of the life cycle of an insect. |
| pycnidium(a) | A flask-shaped asexual fruiting body. Often the imperfect state of Ascomycete fungi. |
| scabrous | Rough to the touch, usually because of the presence of minute prickle-hairs in the epidermis. |
| sclerotium (pl. sclerotia) | In true fungi, a compact mass of hardened tissue with reserve food material, with or without the addition of host tissue or soil, usually having no spores in or on it. |
| sp. | Species (singular). |
| spikelet | The basic unit of the grass inflorescence. In wild rice, composed of a ring of tissue thought to represent the glumes, a lemma, and a palea that surround a male or female (rarely hermaphrodite) flower. |
| spp. | Species (plural). |
| stigma | The top of the ovary or style that receives pollen for effective fertilization. |
| stomata | Apertures or pores in the epidermis bounded by two guard cells; stomata allow the interchange of gases between the atmosphere and the intercellular spaces of the stem. |
| stroma(ata) | A compact mass of fungous tissue on or in which perithecia or pycnidia are produced, often intermingled with tissue of the host or substrate. |
| systematics | Scientific study of the kinds and diversity of organisms and of the relationships between them. |
| taxon | Any taxonomic unit, as species, genus, or tribe. |
| terete | Smooth, well rounded, about cylindrical, but usually tapering at one or both ends. |
| tiller | A lateral shoot of a grass. |
| tineid micromoth | A member of the Tineidae, a group of small moths, commonly called clothes moths. |
| turbidity | Muddiness. |
| vermiform | Worm-like. |

# Index

*Note*: Boldface numbers refer to illustrations and "*t*" refers to tables. In both cases, the number indicates the page on which the material appears, not the figure or table number.

*Aceria tulipae* (mite), 77
Adventitious roots, 11
Air uptake, 40
Alberta, regulations and programs, 96
Algae, 45, 47
Alkalinity (water quality), 41
Ammonium, 42
Anthers, 17, **18**, 22, **28-29**
Anthracnose, 62, **68**
*Apamea apamiformis* (riceworm), 72-74, **75**
*Aphidius obscuripes* (aphid), 77
Aphids, 77
Aquaculture—see Cultivation
Arrowhead, rigid, 44
*Asclepias syriaca* (milkweed), 71
Ascospores, in ergot, 63
Aster leafhopper, 77
Awns, 17, 19, 22, **28-29**

*Bacillus thuringiensis* (bacterial insecticide), 73-74
Bacterial leaf streak, **69**, 71
Beaver dams, 53, 79
Bees, 77
Blackbirds, 79

Bobolinks, 79
*Bombus vagans* (bee), 77
Bract—see Lemma; Palea
*Brasenia* spp. (watershield), 44-45
Bravo (fungicide), 61
Brown spot, 59-61, **66**
  bacterial, **69**, 71
Bulrush, 44
Bur reed, 44-45

Calcium, 42
Canada—see Exports; Federal regulations and programs (Canada); individual provinces
Canoe and flail method—see Harvesting, traditional methods
Carp, 78
Caryopsis, 17, 19, 22
Cattails, 55
Cattle, 79
*Ceratophyllum* spp. (water plants), 45
*Chelonus knabi* (rice stalk borer parasite), 76
*Chilo plejadellus* (pyralid moth; rice stalk borer), 74-76
Chlorotic streaks, **69**

125

Chlorthalanil (fungicide), 61
Chromosome number, 22
*Claviceps* spp. (ergot)
  *purpurea*, 63
  *zizaniae*, 63
*Cochliobolus* spp. (brown spot fungus), 61
  *miyabeanus* 60, 64
  *ativus*, 60, 64
Coleoptile, **12**, 13, 14
*Colletotrichum* spp. (anthracnose), 62, 64
Conidia
  in brown spot, 60
  in ergot, 63
Copper, 42
Cork cell, **15**
Crayfish, 78
*Cricotopus* spp. (midges), 77
Crowsfoot inflorescence, 22
Culm (stem)
  characteristics, 14, 21
  development of, 13
Cultivation, 104–105
  Canadian trends in, 107
  lake production, 9
  paddy production, 9
  —see also Management of natural stands
Culverts (water depth management), 53
Curing, 87, **88**

Dams (water depth management), 53
Deer, 79
Dehulling—see Threshing
Density—see Plant density
Dermestid beetles, 77–78
*Dialictus imitatus* (bee), 77
Diaphragms (in stem), 21
*Dichotomophthoropsis* spp. (brown spot), 60
Dimpler machine, **91**, 92
Dipel (insecticide), 73–74
*Diplodia oryzae* (fungus), 71

Diseases—see Anthracnose; Bacterial leaf streak; Chlorotic streaks; Brown spot; Ergot; Leaf blotch; Leaf sheath and stem rot; Smut; Zonate eyespot
Dithane M45 (fungicide), 61
*Doassansia zizaniae* (fungus), 71
*Donacia* spp. (leaf beetles)
  *aequalis*, 77
  *magnifica*, 77
Dormancy period, 19–20
*Drechslera* spp. (fungus)
  *catenaria*, 71
  *gigantea*, 65, 71
Drying
  current methods, 87–89
  effect on germination, 20, 49–50
  traditional methods, 82
Ducks, 78–79
Duties, 95

*Eleocharis* spp. (spikerush), 44, 55
Embryo, 22
*Entyloma lineatum* (smut), 62, 65
Epiblast, **12**
Epicuticular wax, 14, 16
*Equisetum* spp. (horsetails), 44
Ergot, 63, 70–71
*Eribolus longulus* (wild rice stem maggot), 76–77
Eriophyid mite, 77
*Erysiphe graminis* (fungus), 71
Estuarine wild rice—see *Zizania aquatica* var. *brevis*
Exports (Canada), 95, 96$\underline{t}$

Federal regulations and programs (Canada), 95
Female spikelets—see Spikelets
Fertilizers, 53–54
Flag leaf, 16
Flowers
  description and development, 16–17, **18**
  —see also Panicle; Spikelets
Food value, 92, 94$\underline{t}$

*Fusarium* spp. (brown spot; rot), 60–61, 64

*Gambus bituminosus* (riceworm parasite), 73
Geese, 78–79
Germ, 19
Germination
 dormancy period, 19–20
 effect of drying, 20, 49–50
 forced, 20
 stages in 11, **12**
Glumes (scales), 16–17
 vestigial, 22
Grading, 92
Grains
 anatomy, 19
 development, 19
Grains—see also Harvesting; Processing
Granary weevil, 78
Gravity sorting, 92, **93**

Habitat, 7–8, 39–47
Hard water—see Alkalinity
Harvest
 historical, 80–81
 traditional methods, 81–84
 mechanical, 84–85, **86**
 seasons for, 85, 87
Helminthosporium blight, 60–61
*Helminthosporium* spp.—see *Cochliobolus*
Herbicides, 57
*Hirschmaniella pisquidensis* (nematode), 78
Horsetail, 44
Hover fly, 77
Hull, 19
*Hydrellia* spp. (leafminers)
 *griseola*, 76
 *ischiaca*, 76

Indian use and harvesting methods, 8, 80–84
Insecticides, in riceworm control, 73–74

Interior wild rice—see *Zizania aquatica* var. *interior*
International Wild Rice Council, 106
Internodes, 14, 21
Intraspecific competition—see Weeds
Iron, 42, 48

Lady beetle, 77
Lead, 42–43
Leaf beetles, 77
Leaf blotch, 62, **68**
Leaf sheath and stem rot, 61–62, **67**
Leafminers, 76–77
Leaves
 description, 21–22
 development, 14–16
Lemma, 16–17, 19, 22, **27**
Life cycle, **10**
Light penetration, 49
 effects on tillering, 39–40
Ligule, 16, 21, **28**
Linoleic acid, 92
Linolenic acid, 92
*Lissorhoptrus* spp. (rice water weevils)
 *buchanani*, 77
 *oryzophilus*, 77

*Macrosiphum avenae* (aphid), 77
*Macrosteles fascifrons* (aster leafhopper), 77
*Magnaporthe salvinii* (fungus), 61, 64
Magnesium, 42, 48
Malathion, 73
Male spikelets—see Spikelets
Management of natural stands, 48–58
 lake selection, 48–49
 nutrient levels, 48
 seeding, 49–53
Manchurian water-rice—see *Zizania latifolia*
Mancozeb (fungicide), 61

Manganese, 42
Manitoba
 production figures, 100, 101*t*
 regulations and programs, 95, 97, 100
Marketing, 105–108
Mesocotyl, **12**, 13
Methicarb repellant, 79
Methyl bromide, 20
Midges, 77
Milkweed, 72
Minnesota Wild Rice Growers Association, 105
Moose, 79
Muskrats, 79
*Mycosphaerella zizaniae* (fungus), 71
*Myriophyllum* spp. (water plants), 45

*Nakataea sigmoidea* (fungus), 61
NAWRGA—see Northern Alberta Wild Rice Growers' Association
Nematodes, 78
New Brunswick, 95, 103
New Crop Development program, 95
Nitrogen, 42–43, 48
Nodes, **13**, 14, 21
Nomenclature, 8, 21
Northern Alberta Wild Rice Growers' Association (NAWRGA), 96
Northern wild rice—see *Zizania palustris* var. *palustris*
Nova Scotia, 95, 103
*Nuphar* spp. (water lilies), 44–45, 55, 57
Nutrient value, 92, 94*t*
Nutrients
 recommended concentrations, 48
 seasonal fluctuations in, 41–43
 —see also Fertilizers
*Nymphaea* spp. (water lilies), 44–45, 55, 57

Ontario
 production figures, 100–101, 102*t*
 regulations and programs, 95, 100–101
*Ophiobolus oryzinus* (fungus), 71
*Oryza sativa* (rice), 8
Ovaries, 17, 22
Oxygen levels, 41–42

Paddy culture—see Cultivation
Palea, 16–17, **18**, **19**, **27**
Panicle, 16–19, 22
Papillae, **15**
Parching—see Drying
Pedicels, 22
Pericarp, 19
*Perimegatoma vespulae* (dermestid beetle), 77–78
pH—see Alkalinity
*Phaeoseptoria* spp. (fungus), 62, 64
Phosphorus, 42–43, 48
*Phytobia incisa* (leafminer), 76
Pickerelweed, 44
Plant density, 47, 57–58
Pollen viability, 16
Pollination, 16
*Pontederia cordata* (pickerelweed), 44
Popular names—see Nomenclature
*Potamogeton robbinsii* (underwater plant), 42
Potassium, 42–43, 48
Prince Edward Island, 95, 103
Processing methods, 87–90, **91**, **93**
Production
 effects of water level, 39–40
 fluctuations in, 9
 in Manitoba, 100, 101*t*
 in Ontario, 100–101, 102*t*
 and paddy culture, 104–106
 trends in, 106–107
 in U.S., 104–106
Propiconazol (fungicide), 61

Provincial regulations and
  programs (Canada)—see
  individual provinces
*Pseudomonas syringae* (leaf
  streak), 65, 71
*Ptinus villiger* (spider beetle), 78
Pycnidia, 62, **68**
Pyralid moths, 74, 77

Quality control, 106
Quebec, 95, 103

*Radopholus gracilis* (nematode),
  78
Rhizomes, 21, 55
*Rhopalosiphum* spp. (aphids)
  *niger*, 77
  *nymphaeae*, 77
  *padi*, 77
  *prunifolia*, 77
Rice, 8
Rice stalk borer, 74–76
Rice water weevils, 77
Riceworm, 72–75
Ring effect, 53
Root system, 11, **12**

*Sagittaria rigida* (rigid
  arrowhead), 44
Saskatchewan Indian Agriculture
  Program (SIAP), 97
Saskatchewan
  production trends, **98–99**
  regulations and programs,
    95, 97
Saskatchewan Wild Rice
  Co-operative, 97
Scales—see Glumes; Lemma;
  Palea
Scarification, 90
*Scirpus* spp. (bulrush), 44
Sclerotia
  in ergot, 63, 70–71
  in leaf sheath and stem rot, 61
*Sclerotium* spp. (fungus)
  *hydrophilum*, 61, 64
  *oryzae*, 61

*zizaniae*, 71
Screening, **91**, 92
Seeding, 49–53
  seasons for, 50–53
SIAP—see Saskatchewan Indian
  Agriculture Program
Silica bodies, **15**, 22
*Sitophilus granarius* (granary
  weevil), 78
Smut, 63, **68**
Snapping turtles, 78
Soil quality, 42–43
  management, 53–55
Sora rails, 79
Southern wild rice—see *Zizania
  aquatica* var. *aquatica*
*Sparganium* spp. (bur reed), 44–
  45
Sparrows, 79
Spider beetle, 78
Spikelets
  female, 16–17, 22, **27**, **28–29**
  male, 16–17, 22, **28–29**
Spikerush, 44, 55
Stamens, 17
Stem—see Culm
Sterility
  fungal, 38
  rates and causes, 17
Stigmas, **18**, 22
Stomata, **15**
Stomates, 40
Straw, 53, 57–58
Styles, 22
Sulfate, 41
Sulfur, 42
Syrphid flies, 77

Taxonomic categories, 23–25
Texas wild rice—see *Zizania
  texana*
Thinning, 57–58
Threshing
  current methods, 90, 92
  traditional methods, 82
Tiller roots, 14, 21, **46**, 47
Tillering, induction of, 39–40

Tilt (fungicide), 61
Tineid micromoths, 78
*Toxomerus politus* (fly), 77
*Typha* spp. (cattails), 55

United States
   paddy culture in, 104
   production figures for, 104–106
United Wild Rice (U.S.), 105
*Ustilago esculenta* (fungus), 38

Viability, 17, 19–20

Water chemistry, 41–42
Water depth, 39–40
   management, 53
Water lilies, 44–45, 55, 57
Waterfowl, 78–79
Watershield, 44–45
Weeds, 43–45
   control of, 54–55, 57
Wheat streak mosaic virus, 65, 71
White heads (rice stalk borer), 75–76
Wild rice —see *Zizania*
Wild Rice Harvesting Act (Ontario), 100
Wild rice industry, 105–106
Wild rice stem maggot, 76–77
Winter seeding, 50–53

*Xanthomonas campestris* (leaf streak), 65, 71

Yield per hectare, 87

Yukon, 103

Zinc, 42, 48
*Zizania* spp.
   description, 21–22
   key, 26
   nomenclature, 8, 24
   —see also individual spp. below; Nomenclature
*Zizania aquatica*, **27, 28**
   description, 30, 32, 34
   distribution, **31**
   var. *angustifolia*, 23*t*
   var. *aquatica*, 23*t*
      description, 30
      ergot in, 70
   var. *brevis*, 23*t*, 24
      description, 30, 32
      distribution, **33**
   var. *interior*, 23*t*, 24
      description, 34, 36
   var. *subbrevis*, 24
*Zizania latifolia*, 21, 23*t*, **37**
   chromosome number, 22
   description, 38
*Zizania palustris*, **15**, 22, **27, 29**
   description, 32, 34–36
   distribution, **35**
   var. *interior*, 23*t*
      ergot in, 70
   var. *palustris*, 23*t*
      description, 34
      ergot in, 70
*Zizania texana*, 21, 23*t*, 36–37
Zonate eyespot, **69**, 71

**MOOSE JAW PUBLIC LIBRARY**
633.178 Wil
Wild rice in Canada /

3 3327 00025 3375